SPRINGER TRACTS
IN MODERN PHYSICS

Ergebnisse
der exakten Natur-
wissenschaften

Volume **64**

Editor: G. Höhler

Editorial Board: P. Falk-Vairant S. Flügge J. Hamilton
F. Hund H. Lehmann E. A. Niekisch W. Paul

Springer-Verlag Berlin Heidelberg GmbH 1972

Manuscripts for publication should be adressed to:

G. HÖHLER, Institut für Theoretische Kernphysik der Universität, 75 Karlsruhe 1, Postfach 6380

Proofs and all correspondence concerning papers in the process of publication should be addressed to:

E. A. NIEKISCH, Kernforschungsanlage Jülich, Institut für Technische Physik, 517 Jülich, Postfach 365

ISBN 978-3-662-14957-7   ISBN 978-3-540-37457-2 (eBook)
DOI 10.1007/978-3-540-37457-2

Quasielastic Neutron Scattering for the Investigation of Diffusive Motions in Solids and Liquids

TASSO SPRINGER

Contents

1. Introduction 2

2. Scattering Theory 4
   2.1 Differential Scattering Cross Sections 5
   2.2 Van Hove's Theory 8
   2.3 Asymptotic Behaviour of the Correlation Functions 14
   2.4 Remarks on the Sum Rules 18

3. Methodical and Experimental Aspects 19
   3.1 Frequency and Wave-Number Range; Resolution and Intensity 19
   3.2 Instruments 22
   3.3 Corrections 26
   3.4 Separation of S_{coh} and S_{inc} 27

4. Monoatomic Liquids with Continuous Diffusion 28
   4.1 The Langevin Equation 29
   4.2 Oscillatory Diffusion 31
   4.3 The Memory Function Concept 35
   4.4 Computer Experiments 36
   4.5 Experiments on Liquid Sodium and Argon 38

5. Jump Diffusion in Liquids 41
   5.1 Theoretical Models 41
   5.2 Experiments on Water 45
   5.3 Difference between Continuous and Jump Diffusion 47

6. Diffusion of Hydrogen in Metals 50
   6.1 Theoretical Models 51
   6.2 Experiments 55

7. Rotational Diffusion in Molecular Solids 58
   7.1 Jump Models with Equilibrium Orientations Determined by Crystal
       Symmetry 62
   7.2 Description by Orientational Correlation Functions 64
   7.3 Experiments 69

8. Molecular Liquids 71
   8.1 Remarks on Theory 71
   8.2 Experiments 72

9. Polymeres and other Complicated Systems 76
    9.1 Polymeres 76
    9.2 Different Kinds of Water 79

10. Effects of Coherent Scattering 81
    10.1 Hydrodynamic Description 81
    10.2 Influence of the Liquid Structure 82

11. Quasielastic Scattering and other Methods 87
    11.1 Various Kinds of Scattering Experiments 87
    11.2 Relaxation Methods 90
    11.3 Mössbauer Effect 93

References 94

1. Introduction

If monoenergetic neutrons interact with atomic nuclei tightly bound in a crystal, the energy spectrum of the scattered neutrons consists of two parts: A contribution due to the interaction with the vibrations of the atoms, and an extremely sharp line at energy transfer $\hbar\omega = 0$. Physically, this line corresponds to processes where the neutron has exchanged momentum with the crystal *as a whole* without having suffered any energy exchange with the internal quantum states of the crystal, in analogy to the Mössbauer line for gamma ray interactions. We assume now that the scattering particle is able to perform some translational *diffusion*, as in a liquid or in a solid at higher temperature. This motion leads to a broadening of the line which we call therefore "quasielastic". If there are other degrees of freedom which undergo *random* rather than oscillatory motions (e.g. rotational jumps of molecules in a solid), they also produce a quasielastic line around $\hbar\omega = 0$; this is superimposed on the quasielastic or elastic line discussed before.

By investigating quasielastic scattering one can draw conclusions on the rate and the geometry of such non-periodical motions. For conventional neutron spectrometers with a resolution of the order of 10^{-5} to 10^{-4} eV the observable time scale of the motion is in the region of 10^{-10} to 10^{-12} sec. More recently, a new type of spectrometer has been developped by which this region has been extended to times that are at least two orders of magnitude larger (Section 3). As a consequence, neutron quasielastic scattering now overlaps with the methods used to study relatively slow motions, as nuclear magnetic resonance, ultrasonic and dielectric measurements.

Our review deals with the various applications of quasielastic scattering for problems in physics and chemistry. For each class of problems we first discuss the pertinent theoretical models by which neutron results can be interpreted. This is followed by the description of typical experiments which may serve as examples for the application of these models. We restrict ourselves to investigations concerning *random* motions of atoms and molecules. The investigation of periodical modes will only be treated in so far as they determine the intensity of the quasielastic line, for instance via the Debye-Waller factor. Preferably, we deal with *incoherent* scattering which gives information on the *individual* motion of a particle. Coherent scattering, revealing the effects of cooperative motions of different atoms, will be discussed only briefly: The problems of long-range cooperative fluctuations in liquids, becoming particularly strong near critical points, have been dealt with elsewhere in many articles. As far as *short-range correlations* are concerned this field is not yet developped, neither experimentally nor in theory; only a few qualitative remarks can be made on this subject (Section 10).

Among the subjects to be treated, diffusive motions in *simple liquids* (Section 4 and 5) play the most important role. Quasielastic scattering has been able to reveal how this motion occurs in detail, in particular with regard to the question of its crystalline features. A closely related subject is the diffusion of *hydrogen in metals* (Section 6) which behaves, in certain respects similar to a gas or a liquid. Concerning random *rotations of molecules* in solids (Section 7) the important questions are whether they occur freely, diffusively or jump-wise, and whether they are isotropic in space or not. For *molecular liquids* (Section 8) the interpretation of the experimental results turns out to be more complex than in solids; both rotational and translational motions contribute to the quasielastic line so that the results depend on many parameters. Finally, several problems concerning motions in polymers, macromolecules, and other more complicated systems will be treated (Section 9).

The general *theory* of the experimental method (Section 2), establishing a connection between the atomic motion and the scattered intensity will be worked out in some detail. It will be seen that the treatment of quasielastic scattering is simple because the scattering process can be formulated by means of first Born approximation; furthermore, one has to consider relatively slow atomic motions which can be calculated in terms of classical mechanics. Quantum liquids, like helium will not be treated.

In view of the development of new kinds of spectrometers with higher resolution, in particular at high flux reactors, it is supposed that quasielastic scattering becomes increasingly useful for many fields of research; this holds especially for problems of molecular crystals and other subjects of chemistry. Therefore, quasielastic scattering will develop gradually into a standard method, as, for instance, nuclear magnetic resonance or optical spectroscopy (Section 11). For the reader who wants an introduction into the whole field of neutron spectroscopy we refer to several textbooks [1–3]. In one of them [3] one finds also a description of conventional experimental methods. Further, we draw the reader's attention to the proceedings of the regular international conferences on inelastic neutron scattering [4] which present an overall survey on this field. The proceedings of the last of these conferences, held 1972 in Grenoble, are in print[1].

2. Scattering Theory

The quantity which essentially results from neutron scattering experiments is the *differential scattering cross section* $\mathrm{d}^2\sigma(E_0, E_1, \vartheta)/\mathrm{d}\Omega\mathrm{d}E_1$. It describes the probability with which an incident neutron of energy E_0 is scattered by an angle ϑ into a solid angle element $\mathrm{d}\Omega$, and in an energy intervall between E_1 and $E_1 + \mathrm{d}E_1$ *(Fig. 1)*. In the following, this quantity will be derived using first order perturbation theory.

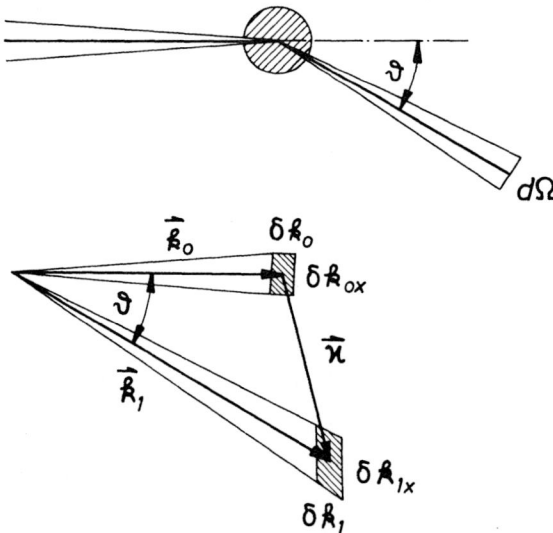

Fig. 1. Neutron scattering process in real and in momentum space. Endpoints of scattering vectors k_0, k_1 distributed over the whole volume of the resolution elements δV_0 and δV_1. The spread δk_0, δk_1 is determined by the energy resolution; δk_{0x}, δk_{1x} by the angular width of the collimators of the spectrometer. (The spread perpendicular to the paper plane is δk_{0y}, δk_{1y}.) Solid lines indicate mean directions. Shaded areas: δV_0, δV_1.

---

[1] The literature for this article has been observed approximately until the middle of 1971. "Letters" have been cited only in a few cases. Publications not specifically dealing with quasielastic neutron scattering, but being related to this subject, have been selected without trying to obtain any completeness.

According to the theory of van Hove [5] it will be shown that the scattering probability is connected with *correlation functions;* they describe the motion of the scattering particles in space and time. Therefore, they allow a calculation of the scattering cross section in terms of simple physical pictures.

2.1 Differential Scattering Cross Sections

The Hamiltonian of the total system, consisting of the neutron and the sample, is written as

$$H = H_0 + (p_n^2/2m) + W. \tag{1}$$

Here H_0 is the Hamiltonian of the undisturbed system, W is the interaction potential between the neutron and the scattering nuclei, $p_n^2/2m$ is the kinetic energy of the neutron, and m its mass. The incident and the scattered waves are characterized by the momentum vectors $\hbar k_0$ and $\hbar k_1$, respectively (k is the wave vector of the neutron, with $k = 2\pi/\lambda$). Denoting the eigenfunctions of the undisturbed system by ψ_n, with the corresponding energies E_n, we have

$$H_0 \psi_n = E_n \psi_n. \tag{2}$$

In first order perturbation theory, the transition probability per unit time of the system from an initial state, k_0, n_0, to a final state, k_1, n_1, is then given by (see for instance [6])

$$w(k_0; n_0 \to k_1; n_1) = 2\pi \hbar^{-1} |\langle k_1 n_1 | W | k_0 n_0 \rangle|^2 \, \delta(E_{n0} - E_{n1} - \hbar\omega). \tag{3}$$

The quantity

$$\hbar\omega = E_0 - E_1 = (\hbar^2/2m)(k_0^2 - k_1^2) \tag{4}$$

is the energy exchanged between the neutron and the sample during the scattering process. The probability for *all* transitions where the neutron wave function changes from k_0 to k_1, is therefore

$$w(k_0 \to k_1) = \sum_{n0} \sum_{n1} p(n_0) \, w(k_0; n_0 \to k_1; n_1). \tag{5}$$

Here $p(n_0)$ is the statistical weight of state n_0 with $\sum p(n_0) = 1$. For a normalized plane wave $L^{-3/2} \exp\{ik_0 r\}$ the incident current density is $j_0 = \hbar k_0/mL^3$ (L^3 = normalization volume). The differential scattering cross section is then defined by the transition rate for a volume element in phase space, dk_1, namely

$$j_0(d^2\sigma/d\Omega \, dE_1) \, d\Omega \, dE_1 \equiv w(k_0 \to k_1)(L/2\pi)^3 \, dk_1. \tag{6}$$

$(L/2\pi)^3 \, \mathrm{d}\boldsymbol{k}_1$ is the number of states with momentum $\hbar \boldsymbol{k}_1$ within $\mathrm{d}\Omega \, \mathrm{d}E_1$. With

$$E_1 = \hbar^2 k_1^2/2m \quad \text{and} \quad \mathrm{d}\boldsymbol{k}_1 = k_1^2 \, \mathrm{d}\Omega \, \mathrm{d}k_1 = mk_1 \, \mathrm{d}E_1 \, \mathrm{d}\Omega/\hbar^2 \tag{7}$$

one finds (Fig. 1)

$$\mathrm{d}^2\sigma/\mathrm{d}\Omega \, \mathrm{d}E_1 = (m/2\pi\hbar^2)^2 \, (k_1/k_0) \, (L^6/N)$$
$$\cdot \sum_{n0} \sum_{n1} p(n_0) \, |\langle \boldsymbol{k}_1; n_1 | W | \boldsymbol{k}_0; n_0 \rangle|^2 \, \delta(E_{n0} - E_{n1} - \hbar\omega). \tag{8}$$

The factor $1/N$ has been included because the cross section is defined per nucleus. In the following, the matrix element will be considered in more detail.

The potential $V(\boldsymbol{r} - \boldsymbol{r}_j)$ of the interaction between a neutron at position $\boldsymbol{r}$ and the nucleus at $\boldsymbol{r}_j$, is unknown. However, because of the very short range ($\sim 10^{-12}$ cm) as compared to the pertinent neutron wave lengths it is allowed to replace the potential by a delta function, writing

$$V(\boldsymbol{r} - \boldsymbol{r}_j) = (2\pi\hbar^2/m) \, a_j \delta(\boldsymbol{r} - \boldsymbol{r}_j). \tag{9}$$

This is the *Fermi pseudo potential* [7]. As a consequence, the detailed properties of the interaction are described by only *one* empirical quantity, a_j. It is called the *scattering amplitude* of the bound nucleus. It can be determined experimentally in various ways, for instance from the cross section of the nucleus at energies where binding effects are negligible, $\sigma_\infty = 4\pi a^2 M^2/(M + m)^2$ (M = atomic mass) or from the refractive index $n = 1 - \lambda^2 N a/2\pi$ (see for instance [1]).

Introducing the effective potential (9) into Eq. (8) and describing the neutron after scattering by a plane wave $L^{-3/2} \exp\{i\boldsymbol{k}_1 \boldsymbol{r}\}$, one obtains

$$\mathrm{d}^2\sigma/\mathrm{d}\Omega \, \mathrm{d}E_1 = (k_1/k_0) \, N^{-1} \sum_{n0} \sum_{n1} p(n_0) \left| \sum_{j=1}^{N} a_j \langle n_1 | \exp\{i\boldsymbol{\kappa}\boldsymbol{r}_j\} | n_0 \rangle \right|^2$$
$$\cdot \delta(E_{n0} - E_{n1} - \hbar\omega)$$
$$= (k_1/k_0) \, N^{-1} \sum_{n0} \sum_{n1} p(n_0) \sum_{i=1}^{N} \sum_{j=1}^{N} a_i a_j \langle n_0 | \exp\{-i\boldsymbol{\kappa}\boldsymbol{r}_i\} | n_1 \rangle$$
$$\cdot \langle n_1 | \exp\{i\boldsymbol{\kappa}\boldsymbol{r}_j\} | n_0 \rangle \, \delta(E_{n0} - E_{n1} - \hbar\omega). \tag{10}$$

Here we have defined the scattering vector

$$\boldsymbol{\kappa} = \boldsymbol{k}_0 - \boldsymbol{k}_1 \tag{11}$$

where

$$\kappa^2 = k_0^2 + k_1^2 - 2k_0 k_1 \cos \vartheta. \tag{12}$$

$\hbar\boldsymbol{\kappa}$ is the momentum which the neutron transfers to the sample during the scattering process. For $\hbar\omega \ll E_0$, the scattering vector depends practically only on ϑ, and not on ω.

In many cases, the sample contains a mixture of isotopes with different scattering amplitudes a_j. We assume that the isotopes are randomly distributed over the positions $\boldsymbol{r}_j$, and introduce the following averages

$$\langle a_i a_j \rangle = \langle a \rangle^2 + \delta_{ij}(\langle a^2 \rangle - \langle a \rangle^2) \tag{13a}$$

with the definitions

$$a_{\text{coh}} \equiv \langle a \rangle = N^{-1} \sum_{j=1}^{N} a_j \quad \text{and} \quad \langle a^2 \rangle \equiv N^{-1} \sum_{j=1}^{N} a_j^2. \tag{13b}$$

Instead of the scattering lengths one often uses the so-called *bound cross sections*[2] which are defined by

$$\sigma_{\text{coh}} = 4\pi\langle a \rangle^2, \quad \sigma_{\text{inc}} = 4\pi(\langle a^2 \rangle - \langle a \rangle^2). \tag{14}$$

The total bound scattering cross section is then $\sigma_s = \sigma_{\text{coh}} + \sigma_{\text{inc}}$ where σ_{inc} is the isotope incoherent cross section.

After having separated the terms $i = j$, the double sum over i and j in (10) can now be written as

$$\begin{aligned} d^2\sigma/d\Omega\,dE_1 = \hbar^{-1}(k_1/k_0)\,[\langle a \rangle^2\, S_{\text{coh}}(\boldsymbol{\kappa}, \omega) \\ + (\langle a^2 \rangle - \langle a \rangle^2)\, S_{\text{inc}}(\boldsymbol{\kappa}, \omega)]. \end{aligned} \tag{15}$$

Here we have introduced the following functions:

$$S_{\text{coh}}(\boldsymbol{\kappa}, \omega) = N^{-1} \sum_{n0,n1} p(n_0) \sum_{i,j=1}^{N} \langle n_0 |\exp\{-i\boldsymbol{\kappa r}_i\}| n_1 \rangle$$
$$\cdot \langle n_1 |\exp\{i\boldsymbol{\kappa r}_j\}| n_0 \rangle\, \delta(E_{n0} - E_{n1} - \hbar\omega). \tag{16a}$$

This is the *coherent scattering law* or the "dynamical structure factor". Furthermore

$$S_{\text{inc}}(\boldsymbol{\kappa}, \omega) = N^{-1} \sum_{n0,n1} p(n_0) \sum_{j} \langle n_1 |\exp\{i\boldsymbol{\kappa r}_j\}| n_0 \rangle^2\, \delta(E_{n0} - E_{n1} - \hbar\omega) \tag{16b}$$

is the *incoherent scattering law*. If all scattering particles are equal, the sum $N^{-1}\sum_j$ and the subscripts j can be dropped. S_{coh} depends on terms which are due to the interference between scattered waves originating

---

[2] Called "bound cross section" because, for a rigidly bound nucleus one would obtain $\int\int (d^2\sigma/d\Omega\,dE_1)\,d\Omega\,dE_1 = \sigma_{\text{coh,inc}}$. This can be shown by writing $G_s(\boldsymbol{r}, t) = \delta(\boldsymbol{r})$ in Eq. (22) below.

from different nuclei $i \neq j$ if these are distinguishable particles; S_{inc}, containing only terms $i = j$, behaves *as if* one had scattering on N isolated nuclei.

So far, it has been tacitly assumed that the spin of the nuclei is $I = 0$. For nuclei with $I \neq 0$, the interaction depends on the orientation between the spin of the neutron and of the nucleus (with scattering lengths a_+ and a_- for parallel and antiparallel spin). For temperatures which are not too low, the nuclear spin orientations are distributed randomly, and no correlation exists with the quantum numbers of the wave functions in space, n_0 and n_1. Under these circumstances one obtains, in analogy to the case of isotope mixtures [2],

$$a_{coh} = \langle a \rangle = [(I+1) a_+ + I a_-]/(2I+1), \tag{17a}$$

$$a_{inc}^2 = \langle a^2 \rangle - \langle a \rangle^2 = I(I+1)(a_+ - a_-)^2/(2I+1) \tag{17b}$$

(spin incoherence). Typical cases, where the above assumptions do not hold, are liquid hydrogen [8] or liquid methane [9]; here the spin quantum numbers are correlated with those of the rotational states.

Eq. (15) includes products of two factors, one depending only on the nuclear interaction, while the other is determined by the dynamical properties and the structure of the atoms in the *undisturbed* sample. Consequently, the scattering law as defined before is of general importance for all kinds of interactions in condensed matter, for instance, S_{coh} for X-ray scattering and for the scattering of light on thermally excited density fluctuations, and S_{inc} for Mössbauer scattering and absorption. Some of these problems will be discussed in Section 11.

It should be pointed out that no experimental evidence exists for any deviation from first order perturbation theory using the Fermi pseudo potential, in spite of the very strong neutron-nucleus interaction. For a discussion of this question we refer to the literature (for instance [1, 10, 11]).

2.2 Van Hove's Theory

The scattering law could be calculated according to (16) if the eigenfunctions n_0 and n_1 of the undisturbed sample are known. This is the case, for instance, in a harmonic crystal, but certainly not for a substance with complicated atomic motions as in liquids. Van Hove [5] has developed an alternative representation of the matrix element formula in terms of the time-dependent atomic coordinates. The derivation will not be given in full length; however, its main features will be treated subsequently.

At first, space vectors describing the motion of the scattering nucleus are replaced by time dependent Heisenberg operators[3]

$$r_j(t) = \exp\{iH_0 t/\hbar\}\, r_j \exp\{-iH_0 t/\hbar\} \,. \tag{18}$$

Introducing now the Fourier representation of the δ-function describing energy conservation, into (16)

$$\delta(E_{n0} - E_{n1} - \hbar\omega) = (2\pi\hbar)^{-1} \int\limits_{-\infty}^{+\infty} dt \exp\{it(E_{n0} - E_{n1} - \hbar\omega)\}$$

and using the completeness of the system, $\Sigma |n\rangle \langle n| = 1$, the scattering law can be written as

$$S_{\mathrm{coh}}(\boldsymbol{\kappa}, \omega) = (2\pi\hbar)^{-1} \int I(\boldsymbol{\kappa}, t) \exp\{-i\omega t\}\, dt \,, \tag{19a}$$

$$S_{\mathrm{inc}}(\boldsymbol{\kappa}, \omega) = (2\pi\hbar)^{-1} \int I_s(\boldsymbol{\kappa}, t) \exp\{-i\omega t\}\, dt \,. \tag{19b}$$

I and I_s are the *intermediate scattering functions*, defined by

$$I(\boldsymbol{\kappa}, t) = N^{-1} \sum_{i=1}^{N} \sum_{j=1}^{N} \langle \exp\{-i\boldsymbol{\kappa} r_i(0)\} \exp\{i\boldsymbol{\kappa} r_j(t)\} \rangle \,, \tag{20a}$$

$$I_s(\boldsymbol{\kappa}, t) = N^{-1} \sum_{i=1}^{N} \langle \exp\{-i\boldsymbol{\kappa} r_i(0)\} \exp\{i\boldsymbol{\kappa} r_i(t)\} \rangle \,. \tag{20b}$$

$\langle\ldots\rangle$ means the thermal average of the expectation value of the operator enclosed in the brackets. Following van Hove's treatment we define now correlation functions

$$G(\boldsymbol{r}, t) = (2\pi)^{-3} \int \exp\{-i(\boldsymbol{\kappa} \boldsymbol{r} - \omega t)\}\, S_{\mathrm{coh}}(\boldsymbol{\kappa}, \omega)\, d\boldsymbol{\kappa} d\omega \,, \tag{21a}$$

$$G_s(\boldsymbol{r}, t) = (2\pi)^{-3} \int \exp\{-i(\boldsymbol{\kappa} \boldsymbol{r} - \omega t)\}\, S_{\mathrm{inc}}(\boldsymbol{\kappa}, \omega)\, d\boldsymbol{\kappa} d\omega \,. \tag{21b}$$

From these equations one obtains immediately

$$S_{\mathrm{coh}}(\boldsymbol{\kappa}, \omega) = (2\pi)^{-1} \int \exp\{i(\boldsymbol{\kappa} \boldsymbol{r} - \omega t)\}\, G(\boldsymbol{r}, t)\, d\boldsymbol{r} dt \,, \tag{22a}$$

$$S_{\mathrm{inc}}(\boldsymbol{\kappa}, \omega) = (2\pi)^{-1} \int \exp\{i(\boldsymbol{\kappa} \boldsymbol{r} - \omega t)\}\, G_s(\boldsymbol{r}, t)\, d\boldsymbol{r} dt \,. \tag{22b}$$

Combining (22) with the intermediate scattering functions (20), and by making use of the relation $\delta(\boldsymbol{r}) = (2\pi)^{-3} \int d\boldsymbol{\kappa} \exp\{i\boldsymbol{\kappa} \boldsymbol{r}\}$, one finds (see [2], p. 517)

$$G(\boldsymbol{r}, t) = N^{-1} \left\langle \sum_{i=1}^{N} \sum_{j=1}^{N} \int d\boldsymbol{r}' \delta[\boldsymbol{r} + r_i(0) - \boldsymbol{r}'] \delta[\boldsymbol{r}' - r_j(t)] \right\rangle. \tag{23}$$

For distinguishable particles it is useful to separate (23) into a *self* and a *distinct* part, namely

$$G(\boldsymbol{r}, t) = G_s(\boldsymbol{r}, t) + G_d(\boldsymbol{r}, t) \tag{24}$$

---

[3] So far, we deal with quantum mechanical theory. Later on, we consider classical systems only. In such cases, the $r_j(t)$ are ordinary time-dependent vectors in space.

where G_s and G_d have the following meaning

$$G_s(r, t) = N^{-1} \left\langle \sum_{i=1}^{N} \int dr' \delta[r + r_i(0) - r'] \delta[r' - r_i(t)] \right\rangle, \qquad (25a)$$

$$G_d(r, t) = N^{-1} \left\langle \sum_{\substack{i=1 \\ i \neq j}}^{N} \sum_{j=1}^{N} \int dr' \delta[r + r_i(0) - r'] \delta[r' - r_j(t)] \right\rangle. \qquad (25b)$$

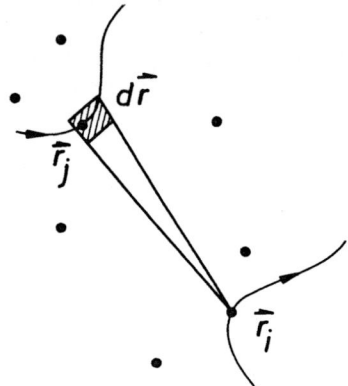

Fig. 2. For the interpretation of the static pair correlation function $G_d(r, 0) = g(r)$ in Eq. (26). Dots indicate the centers of atoms

These functions can be understood in the following way. For $t = 0$, the operators commute and the integration over the product of δ-functions can be carried out, giving

$$G(r, 0) = \delta(r) + N^{-1} \left\langle \sum_{\substack{i \neq j}}^{N} \sum^{N} \delta[r + r_i(0) - r_j(0)] \right\rangle \equiv \delta(r) + g(r). \qquad (26)$$

Here the self-part $\delta(r)$ results from the terms $i = j$. The distinct part is the *static pair correlation function*. To interpret the distinct part we consider a vector r, which originates at a certain particle coordinate, say r_i, and whose endpoint is in a small volume element dr around r_j. Then, as can be seen from *Fig. 2*, the double sum in (26), integrated over this volume element, counts the number of such pairs of particles at r_i and r_j which are simultaneously separated by $r = r_i - r_j$. Consequently, $g(r) \, dr$ according to (26) is the *ensemble averaged probability of finding a particle in a volume element at r, if another one is, simultaneously, at $r = 0$.*

Obviously, the scattering law integrated over the energy transfer is just the Fourier transform of the function $g(r)$, because the integration over ω transforms $\exp\{-i\omega t\}$ into $\delta(t)$, therefore, from (22)

$$S(\boldsymbol{\kappa}) = \int_{-\infty}^{\infty} S(\boldsymbol{\kappa}, \omega)\, d\omega = \int d\boldsymbol{r} \exp\{i\boldsymbol{\kappa}\boldsymbol{r}\}\, G(\boldsymbol{r}, 0)$$
$$= 1 + \int \exp\{i\boldsymbol{\kappa}\boldsymbol{r}\}\, g(\boldsymbol{r})\, d\boldsymbol{r}\,. \tag{27}$$

$S(\boldsymbol{\kappa})$ is the *structure factor*, well-known in X-ray diffraction, where the ω-integration is properly performed because $\hbar\omega$ is very small compared to the incident energy.

For $t > 0$, $G(\boldsymbol{r}, t)$ is a complex function without a classical meaning (for a discussion of the imaginary part of $G(\boldsymbol{r}, t)$ see [1, 13]). If, however, the system behaves as if it were *classical*, the operators commute and the integration (25) can be carried out as before, leading to real functions, namely

$$G_s^{(cl)}(\boldsymbol{r}, t) = N^{-1} \left\langle \sum_{i=1}^{N} \delta[\boldsymbol{r} + \boldsymbol{r}_i(0) - \boldsymbol{r}_i(t)] \right\rangle \tag{28a}$$

for the self part, and

$$G_d^{(cl)}(\boldsymbol{r}, t) = N^{-1} \left\langle \sum_{\substack{i=1 \\ i \neq j}}^{N} \sum_{j=1}^{N} \delta[\boldsymbol{r} + \boldsymbol{r}_i(0) - \boldsymbol{r}_j(t)] \right\rangle \tag{28b}$$

for the distinct part. In analogy to (26), these functions have the following meaning:

Assume that, at an arbitrary time t' and position $\boldsymbol{r}'$ there is a particle. Then $G_d(\boldsymbol{r} - \boldsymbol{r}', t - t')\, d\boldsymbol{r}$ is the ensemble averaged probability of finding another particle in $d\boldsymbol{r}$ at a distant position $\boldsymbol{r}$ for a later time t. In analogy, $G_s(\boldsymbol{r} - \boldsymbol{r}', t - t')\, d\boldsymbol{r}$ is the probability of finding the *same* particle at $\boldsymbol{r}$ for a time t. These are the so-called time-dependent *pair* and *self correlation functions*.

Fig. 3 shows the behaviour of these functions for atoms in a liquid. Because, per definition, the particle must be at $\boldsymbol{r} = 0$ for $t = 0$, one has

$$G_s(\boldsymbol{r}, 0) = \delta(\boldsymbol{r})\,. \tag{29}$$

With increasing time, this delta function broadens continuously due to the diffusive motion of the atoms. $G_d(\boldsymbol{r}, t)$ reveals pronounced maxima which are due to short-range order existent in a liquid. With increasing time, these correlations get lost; after several 10^{-12} sec in a typical liquid, $G(\boldsymbol{r}, t)$ approaches $\bar{\varrho}$, the average density. The asymptotic behaviour of G_s and G_d will be discussed in more general terms in Section 2.3.

In the following, $G_s(\boldsymbol{r}, t)$ will be very often represented by the so-called *Gaussian approximation* [15–18], namely

$$G_s(\boldsymbol{r}, t) = [4\pi\gamma(t)]^{-3/2} \exp\{-r^2/4\gamma(t)\}\,. \tag{30}$$

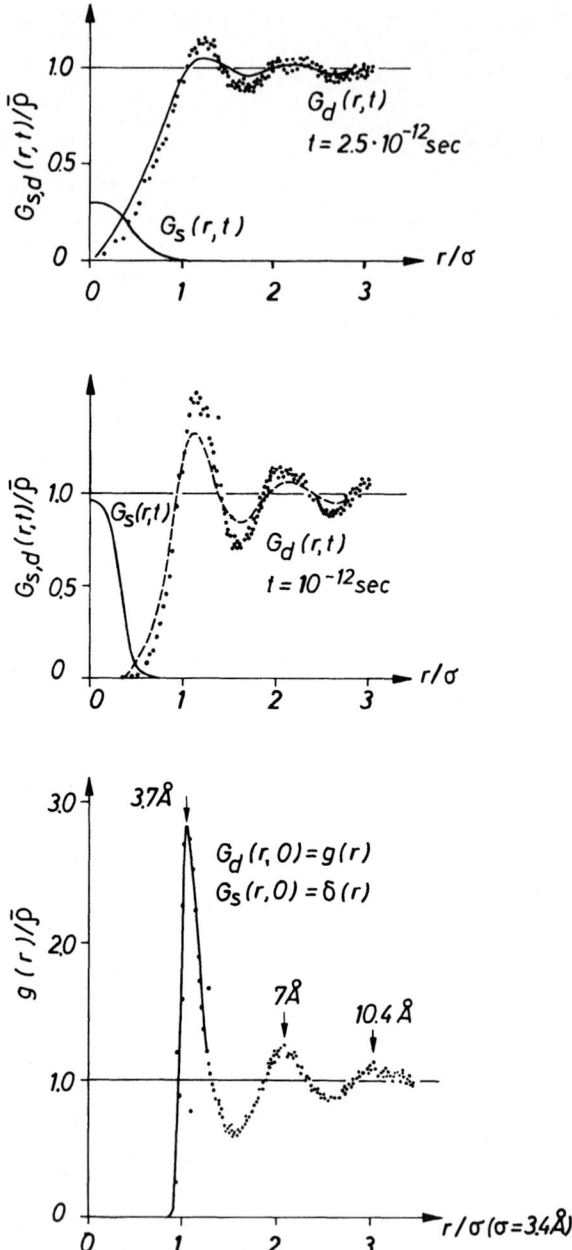

Fig. 3. Self and distinct correlation functions, G_s and G_d. Points: G_d according to computer experiments; from Rahman [14]. Dashed: Convolution approximation (Section 10.2). G_s (solid line) is given schematically. $G_s(r, 0) = \delta(r)$; $G_d(r, t \to \infty) = \bar{\varrho}$ = average density

For a classical system, $\gamma(t)$ is the mean square deviation of a particle which has started from some origin $r = 0$ at $t = 0$, namely

$$\langle r^2(t) \rangle = \int r^2 G_s(r, t)\, dr = 6\gamma(t) . \tag{31}$$

Thus, in the frame of Eq. (30), $S_{\text{inc}}(\kappa, \omega)$ is determined completely if the mean square displacement of the scattering nucleus is known.

The Gaussian form (30) can be shown to hold rigorously for an ideal gas, for a harmonic crystal, and for liquid diffusion at large times [16]. Furthermore, it holds rigorously in any case for $t \to 0$ because then the atoms behave as if they were free. It has been shown that, for atoms in a liquid, considerable deviations from (30) appear at intermediate times [14, 18, 19]. Except for small times, the Gaussian formula obviously does not hold for a particle diffusing in a *solid*. Here it stays mainly on discrete positions determined by the lattice (Section 6); the same statement holds for scattering on particles bound to rotating molecules (Section 7).

All theories to be discussed later make use of the classical correlation functions, defined by (28). To justify this approximation [5], we consider a liquid at temperature T with atoms of mass M, where the average de Broglie wave length is given by

$$\lambda_M \simeq \hbar (2 M k_B T)^{-1/2} . \tag{32}$$

λ_M is of the order of $10^{-9} \ldots 10^{-8}$ cm. Evidently, it makes no sense to calculate G_s classically for times where the average square radius of the probability distribution $\langle r^2(t) \rangle$ in Fig. 3 is comparable with, or smaller than λ_M^2. Because, for small times, we can write $\langle r^2 \rangle = (k_B T / 2M)\, t^2$ (Eq. (68)) the classical description of G_s is expected to fail as soon as $\langle r^2 \rangle \lesssim \lambda_M^2$ or

$$t \lesssim \hbar / k_B T \simeq \ldots 10^{-13} \text{ sec} . \tag{33}$$

For most of the problems to be discussed later, the characteristic times between successive steps of the diffusive motion are longer than 10^{-12} sec. Consequently, the classical description will be justified in this sense, and the subscript "cl" will be ommitted from now on.

According to (22) a real-valued classical $G(r, t)$ necessarily leads to a $S(\kappa, \omega)$ being even in ω. This violates the principle of detailed balance according to which $S(\kappa, \omega)$ must fulfill the condition (for instance [20])

$$S(\kappa, \omega) = \exp\{\hbar\omega/2k_B T\}\, S_0(\kappa, \omega) \tag{34}$$

where $S_0(\kappa, \omega)$ is even in ω. By introducing a new time variable $t + i\hbar/2k_B T$ instead of t into the correct non-classical correlation function one obtains a function

$$G_s^0(r, t) = G_s(r, t + i\hbar/2k_B T) \tag{35}$$

which is real and the Fourier transform of $S_0(\kappa, \omega)$ [18]. Thus, a time shift $i\hbar/2k_B T$ makes the quantum mechanical function real, whereas a shift in the opposite sense "improves", to first order, a function calculated in terms fo classical physics. The term $i\hbar/2k_B T$ appears in all quantum mechanical calculations.

The *classical self-correlation function* G_s can be obtained in various ways. One method is to solve the equation of motion of the diffusing particle under the influence of a random force which replaces, in a summary way, the interaction with the surrounding atoms [21]. From the first integration of this so-called Langevin-equation one obtains easily the *auto-correlation function of the velocity*, namely

$$\psi(t) = \langle \boldsymbol{v}(t') \, \boldsymbol{v}(t' + t) \rangle / \langle \boldsymbol{v}^2 \rangle \,. \tag{36}$$

In Eq. (36), $\boldsymbol{v}(t')$ and $\boldsymbol{v}(t' + t)$ are the velocity vectors of the same particle at a time t' and at a later time $t' + t$ where $\langle ... \rangle$ means an ensemble average which does not depend on the time origin t'.

Instead of this function it is often convenient to use its Fourier transform, which, apart from a temperature dependent factor, is proportional to the socalled *"generalized frequency spectrum"* $f(\omega)$. For a harmonic crystal, it is identical with the familiar density of states [15, 52].

Within the Gaussian approximation (30) one can easily calculate the intermediate scattering function (20b) which is

$$I_s(\boldsymbol{\kappa}, t) = \exp\{-\kappa^2 \gamma(t)\} \,. \tag{37}$$

The square width $\gamma(t)$ of the Gaussian can be directly connected with the velocity auto-correlation function by means of [15]

$$\gamma(t) = \int_0^t dt_1 \int_0^{t_1} dt_2 \langle v_\kappa(t_2) \, v_\kappa(t_1) \rangle \tag{38}$$

where v_κ is the velocity component along $\boldsymbol{\kappa}$. Eq. (38) is equivalent to

$$\gamma(t) = (1/3) \int_0^t (t - t_1) \langle \boldsymbol{v}(0) \, \boldsymbol{v}(t_1) \rangle \, dt_1 \tag{39}$$

for an isotropic system. This approach to determine $G_s(\boldsymbol{r}, t)$ will be used throughout Section 4.

An alternative procedure is to treat the diffusion process by a rate equation which has $G_s(\boldsymbol{r}, t)$ as a solution if the initial conditions are chosen properly. In this description, only the geometry and the rate of diffusive jumps enters, and all details of the diffusive steps are neglected. This method will be applied to describe diffusive motions in solids (Section 6 and 7). In a certain sense, the most rigorous method is the solution of the equations of motion for a number of 10^2 to 10^3 interacting particles by means of "computer experiments". From these one can extract the correlation functions G_s, G_d, as well as $\gamma(t)$ or $\psi(t)$ (Section 4.4).

2.3 Asymptotic Behaviour of the Correlation Functions

The correlation functions can be formulated in terms of the microscopic space and time dependent particle density. According to its definition, it indicates the position of each particle by a delta function, namely

$$\varrho(\boldsymbol{r}, t) = \sum_{i=1}^N \delta[\boldsymbol{r} - \boldsymbol{r}_i(t)] \,. \tag{40}$$

If this function is introduced into (23) one easily obtains

$$G(r, t) = N^{-1} \int \langle \varrho(r' - r, 0) \varrho(r', t) \rangle \, dr' \qquad (41)$$

so that $G(r, t)$ can be considered as the auto-correlation function of the particle density, averaged over the sample volume. For large distances in space and time, the microscopic density functions in (41) are statistically independent from each other so that the average of the product approaches the product of the averages. This leads to

$$G^\infty(r) = \lim_{r, t \to \infty} G(r, t) = N^{-1} \int \overline{\varrho}(r' - r) \, \overline{\varrho}(r') \, dr' \,. \qquad (42)$$

Here the time-independent average density is given by

$$\overline{\varrho}(r) = \left\langle \sum_{i=1}^{N} \delta[r - r_i(t)] \right\rangle = \langle \varrho(r, t) \rangle \,. \qquad (43)$$

For a crystal, $\overline{\varrho}(r)$ and therefore $G^\infty(r)$ is periodic in space. On the other hand, for a homogenous system like a liquid, the integration in (41) can be carried out and one gets

$$G(r, t) = \langle \varrho(0, 0) \varrho(r, t) \rangle / \overline{\varrho} \,. \qquad (44)$$

Here $G^\infty(r) = \overline{\varrho} = N/V$ is the density of the liquid (V means the sample volume).

In an analogous way we can factorize the self-correlation function $G_s(r, t)$ in (25) which results in

$$G_s^\infty(r) = N^{-1} \sum_{i=1}^{N} \int \langle \delta[r + r_i(0) - r'] \rangle \langle [\delta(r' - r_i(t)] \rangle \, dr' \,. \qquad (45)$$

In the following, we introduce the time-independent average probability per unit volume of finding the particle i at r', namely

$$p_i(r') = \langle \delta[r' - r_i(t)] \rangle \qquad (46)$$

and we get

$$G_s^\infty(r) = N^{-1} \sum_{i=1}^{N} \int p_i(r' - r) \, p_i(r') \, dr' \,. \qquad (47)$$

Assume, first, that the scattering particles are able to diffuse freely throughout the whole sample volume V. Then, for $t \to \infty$, the probability p_i and $G_s^\infty(r)$ approache $1/V$, so that $G_s^\infty(r) \to 0$ for $V \to \infty$.

Assume, on the other hand, that the motion of each individual particle is *restricted to a finite volume*, for instance if the scattering nucleus oscillates or rotates around a fixed equilibrium position. Then p_i and $G_s^\infty(r)$ are finite. Under the assumption that all N scattering particles are equivalent, one can drop the index i and carry out the summation in (47). This leads to

$$G_s^\infty(r) = \int p(r' - r)\, p(r')\, dr' .\tag{48}$$

We now subdivide $G_s(r, t)$ in the stationary part $G_s^\infty(r)$ and in a part which decays to zero for $t \to \infty$, namely

$$G_s'(r, t) = G_s(r, t) - G_s^\infty(r) .\tag{49}$$

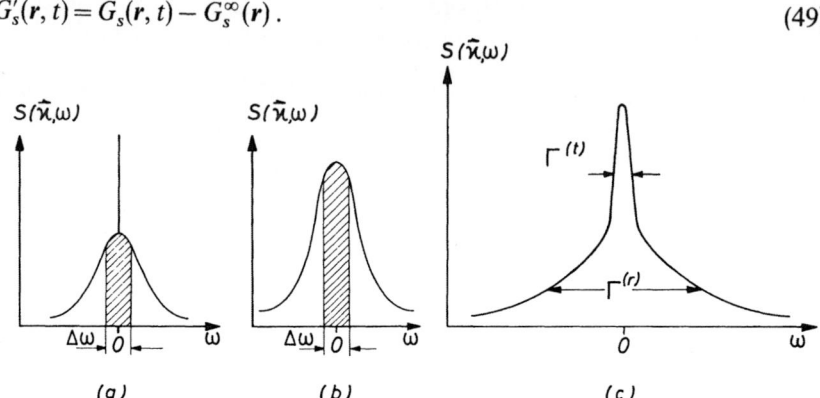

(a)   *(b)*   *(c)*

Fig. 4. a Composite neutron spectrum, containing a sharp elastic line in addition to the quasielastic line. This occurs if the scattering particle performs some random motion and is found within a finite volume *for all times* so that $G_s(r, \infty)$ is finite. The measured intensity I_0 within the resolution interval (shaded) remains finite if $\Delta\omega$ goes to zero (see Section 2.3). b Quasielastic spectrum for neutrons scattered on a particle whose motion is not restricted in space, so that $G_s(r, t) \to 0$ if $t \to \infty$. Consequently, I_0 goes to zero for $\Delta\omega \to 0$. c Composite spectrum for a scattering particle performing a translational diffusive *and* some random rotational motion, e.g. a proton in a molecular liquid (Section 8)

Obviously, the time-dependent and decaying term is responsible for scattering with finite energy transfer. This results in quasielastic and inelastic scattering. On the other hand, the Fourier transform of the stationary distribution gives rise to a *sharp elastic line*, like the Mössbauer line *(Fig. 4a)* with an intensity[4]

$$S_{inc}^{el}(\kappa) = \int \exp\{-i\kappa r\}\, G_s^\infty(r)\, dr$$
$$= \int\int \exp\{-i\kappa r\}\, p(r - r')\, p(r')\, dr'\, dr .\tag{50}$$

According to a well-known theorem of Fourier transformation this is equal to

$$S_{inc}^{el}(\kappa) = |\int \exp\{-i\kappa r\}\, p(r)\, dr|^2 .\tag{51}$$

---

[4] Our introduction of an "incoherent structure factor" follows reference [1], p. 105 ff. — The existence of a sharp elastic line for light scattering on atoms enclosed in a box of finite size has been discussed by Dicke [22].

This function will be called *incoherent structure factor*, in analogy to the coherent structure factor in Eq. (54) below. We emphasize that the intensity of the purely elastic line, $S_{inc}^{el}(\kappa)$, is determined by the diffraction of the neutron wave on the probability distribution in space of an *individual* scattering particle, being spread over a finite volume; this may occur by virtue of vibrational, rotational, tunneling, or other motions ("thermal cloud" of the particle). This must not be confused with *coherent* scattering taking place on *different* particles distributed in space, as it is described by the coherent structure factor

$$S_{coh}^{el}(\kappa) = |\int \exp\{-i\kappa r\} \varrho(r)\,dr|^2 . \tag{52}$$

We try to make these equations plausible in more detail by regarding an experiment where incoherent scattering takes place on a freely diffusing particle whose motion is described by $G_s(r, t)$. Assume that the monochromator and the analyzer of the neutron spectrometer are positioned at $E \simeq E_0$, i.e. at $\hbar\omega \simeq 0$, and that its resolution function is, for simplicity, rectangular-shaped with a width $\Delta\omega$. Then, the intensity I_0 scattered into this resolution interval can be calculated from (22), giving

$$I_0(\kappa, \Delta\omega) \sim \int_{-\Delta\omega/2}^{\Delta\omega/2} S_{inc}(\kappa, \omega)\,d\omega \simeq (2\pi\hbar)^{-1} \int d r \exp\{i\kappa r\} \int G_s(r, t)$$
$$\cdot \Delta\omega \frac{\sin(t\Delta\omega/2)}{t\Delta\omega/2}\,dt . \tag{53}$$

According to the properties of the window function $\sin x/x$, this is an average of $G_s(r, t)$ over a time interval of the order of $1/\Delta\omega$. We consider this as the "coherence time" of the neutron wave in the scattering process; it is determined by the choice of instrumental resolution.

If we now reduce the resolution width $\Delta\omega$ to smaller and smaller values, the average extends over larger and larger times. For a freely diffusing particle, $G_s(r, t)$ goes to zero for $t \to \infty$. Therefore, the measured intensity will also vanish (shaded area in *Fig. 4b*).

If, on the other hand, the particle performs a motion bound to a finite volume, $G_s(r, t)$ is finite for $t \to \infty$ and $I_0(\Delta\omega)$ remains finite, how small the resolution ever will be (Fig. 4a). This means that the scattering law contains a δ-function as explained before. To be able to observe the incoherent structure factor experimentally, the resolution width must fulfill the requirement $\Delta\omega \lesssim 1/\tau$, where τ is a characteristic time of the motion of the scattering particle around its equilibrium site. For a thermally excited jump motion, $1/\tau$ would be the jump rate. (For the case of a tunneling process see [12].)

Let us assume we had calculated the *incoherent* structure factor $S_{\text{inc}}^{\text{el}}(\kappa)$ for a single proton bound to a molecule performing some jump rotation between different quasi-equilibrium sites in an orientationally disordered crystal. Let us further assume we could replace this proton by a *coherently* scattering atom and neglect scattering from the other atoms in the molecule. The asymptotic distribution $p(r)$ of the single proton in one cell would then be identical to the distribution $\langle \varrho(r) \rangle_{\text{orient}}$ of the coherently scattering atom, averaged over all cells of the disordered crystal, or over all possible orientations the molecule can occupy. As a consequence, $S_{\text{inc}}^{\text{el}}(\kappa)$ in Eq. (51) would then be identical to the coherent structure factor for elastic diffraction on this crystal,

$$S_{\text{coh}}^{\text{el}}(\kappa) = (8\pi^3/V_0) \,|\textstyle\int \exp\{-i\kappa r\} \, \langle \varrho(r) \rangle_{\text{orient}} \, dr|^2 \,. \tag{54}$$

Contrary to the incoherent structure factor (51), Eq. (54) is only defined if κ is equal to a reciprocal lattice vector. The essential difference is that the neutron spectrum reveals the *dynamics* of the disorder which can not be seen in the coherent diffraction pattern of X-rays.

2.4 Remarks on the Sum Rules

The scattering law fulfills certain sum rules which lead to the so-called *moments*

$$\langle \omega^{2n} \rangle = \int\limits_{-\infty}^{+\infty} S(\kappa, \omega) \, \omega^{2n} d\omega \Big/ \int\limits_{-\infty}^{+\infty} S(\kappa, \omega) \, d\omega \,. \tag{55}$$

These can be rigorously derived for classical [23] as well as for quantum mechanical systems [24, 15], for instance $\int S_{\text{inc}}(\kappa, \omega) d\omega = 1$ for the integrated scattering law, and, for a classical system,

$$\langle \omega^2 \rangle_{\text{inc}} = \kappa^2 k_B T/M \,; \quad \langle \omega^2 \rangle_{\text{coh}} = \langle \omega^2 \rangle_{\text{inc}}/S(\kappa) \,.$$

One further gets the quantity

$$\langle \omega^4 \rangle_{\text{inc}}/3\langle \omega^2 \rangle_{\text{inc}}^2 = 1 + \Omega_0^2/\langle \omega^2 \rangle_{\text{inc}} \tag{56}$$

where the characteristic (average) frequency of the atoms is defined as

$$\Omega_0^2 = (4\pi/3M) \int g(r) \, (\partial^2 V(r)/\partial x^2) \, dr \,. \tag{57}$$

For an ideal gas, the interaction potential $V(r)$ is zero and the ratio (56) is unity which corresponds to a *Gaussian*-shaped spectrum. With increasing strength of $V(r)$ and with decreasing κ, the ratio increases which indicates a *contraction* of the spectrum near $\hbar\omega = 0$ and, at the same time, an enhancement of the wings. Thus the spectrum becomes more like a *Lorentzian* (see Fig. 17 in Section 5.3).

Fig. 5 shows $\langle \omega^4 \rangle/3\langle \omega^2 \rangle^2$ for incoherent and also for coherent scattering. In the latter case, the liquid structure is responsible for pronounced oscillations (see Section 10.2). A reliable experimental determination of the higher moments [25] which would be of great physical value, is difficult. A comparison of the lower moments with their known theoretical values could be used to check the reliability of the experimental results.

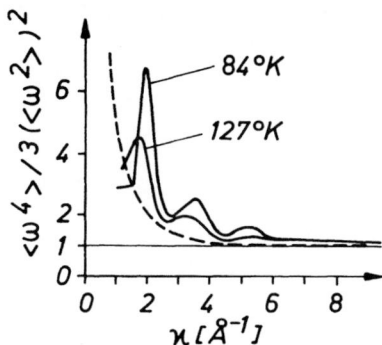

Fig. 5. Solid lines: Ratio of 4th and square of 2nd moment of the scattering law for coherent scattering on liquid argon as a function of scattering vector (from de Gennes [23]). Dashed: Incoherently scattering liquid (qualitatively). Maxima indicate "line narrowing" (Section 10.2). The asymptotic values for large κ correspond to a Gaussian-shaped spectrum

3. Methodical and Experimental Aspects

Before entering the discussion of the various classes of problems and experiments we describe certain methical aspects of neutron spectroscopy in so far as they concern small energy transfers. It will be shown that the central problem in all these experiments is the restriction of instrumental resolution due to the available luminosity of the sources. Therefore, we will explain the systematic connection between intensity and resolution in more detail. Because of this intensity shortage, in most scattering experiments the resolution width has to be chosen as large or even larger than the quasielastic width to be studied. Consequently, very careful resolution corrections play an important role in such experiments. Furthermore, in this section we will deal with more recent developments in the field of high resolution neutron spectroscopy.

3.1 Frequency and Wave-Number Range; Resolution and Intensity

As has been demonstrated in Eq. (22), the neutron scattering experiment performs a Fourier analysis of the correlation functions $G_s(r, t)$ or $G_d(r, t)$ in space and time; the frequency $\omega/2\pi$ and the wave vector κ

of the Fourier components are defined by energy and momentum transfer during the scattering process, $\hbar\omega = E_0 - E_1$ and $\hbar\boldsymbol{\kappa} = \hbar(\boldsymbol{k}_0 - \boldsymbol{k}_1)$, respectively. According to the properties of Fourier integrals, the behaviour of the original function at large arguments is mainly responsible for the transformed function at small arguments, and vice versa. As a consequence, the scattering intensity measured at small (large) momentum and energy transfers depends essentially on the behaviour of the correlation functions at large (short) distances and times.

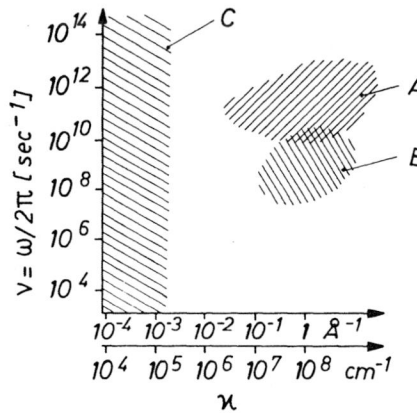

Fig. 6. Range of frequency $\nu = \omega/2\pi$ and wave vector κ of the Fourier components accessible to scattering experiments. A: Region of conventional neutron spectrometers. B: region of the "90°-spectrometer" (Section 3.2). C: Raman, infrared, and Laser spectroscopy

Fig. 6 shows the regions of a κ, ω-plane which are accessible to neutron spectroscopy. Region A corresponds to conventional time-of-flight and triple axis spectrometers working in an energy region of a few 10^{-3} to 10^{-2} eV, and in a κ-region between a few $0{,}01 \text{ Å}^{-1}$ and several Å^{-1}. Typical energy resolutions are in the range of about $\delta\hbar\omega \simeq 0{,}5 \cdot 10^{-4}$ to $5 \cdot 10^{-4}$ eV. The high resolution region B corresponds to the "90° spectrometer" which will be described later; it has a resolution of about $\delta\hbar\omega \simeq 5 \cdot 10^{-7}$ eV. Optical spectroscopy covers, since the availability of Laser spectrometers, an enormous ω-region, whereas the κ-region is necessarily restricted to $\kappa < 10^{-3} \text{ Å}^{-1}$.

The classical relaxation methods, like nuclear magnetic resonance, ultrasonic and dielectric measurements apply to a range in general below 10^{10} sec^{-1}. With X-rays, a κ-region can be covered similar to that of neutrons. However, no means exists for an energy analysis in the range which is of interest for the problems to be discussed here.

As stated before, the resolution to be obtained in neutron spectroscopy is mainly determined by the *intensity* of the sources, and not by the properties of the instruments as such. To understand this on a quantitative basis, we calculate the intensity at the detector of a typical neutron spectrometer. The volume elements in momentum space, as defined by the monochromator and the analyzing systems of the spectrometer are given by the following expressions (see Fig. 1)

$$\delta V_0 = \delta k_0 \delta k_{0x} \delta k_{0y} \quad \text{and} \quad \delta V_1 = \delta k_1 \delta k_{1x} \delta k_{1y} . \tag{58}$$

The longitudinal components, δk_0 and δk_1, are determined by the resolution width in wave-number, namely

$$\delta k_0 = 2k_0 \delta E_0/E_0 \quad \text{and} \quad \delta k_1 = 2k_1 \delta E_1/E_1 . \tag{59}$$

δE_0 and δE_1 are the energy resolution widths of the spectrometer for the incident and the scattered beam, respectively. The transversal components $\delta k_{0x}, \delta k_{0y}$, etc. are defined by the horizontal and the vertical angular spread of the neutron beams. As has been demonstrated by Maier-Leibnitz [26] the count rate Z at the detector of a neutron spectrometer is connected with these quantities in a very simple way; Z is directly proportional to the total thermal reactor flux, to the scattering law $S(\kappa, \omega)$, to the area of the sample and of the detector, and to the product of the two volume elements (58), namely

$$Z \sim \delta V_0 \delta V_1 = \delta k_0 \delta k_{0x} \delta k_{0y} \delta k_1 \delta k_{1x} \delta k_{1y} . \tag{60}$$

We consider now the square width of the resolution function of the energy transfer; it is given by

$$\langle \delta(\hbar\omega^2) \rangle = \langle \delta E_0^2 \rangle + \langle \delta E_1^2 \rangle \tag{61}$$

where $\langle \delta E_0^2 \rangle$ and $\langle \delta E_1^2 \rangle$ are the mean square widths of the distributions of δE_0 and δE_1. For an optimum layout of the spectrometer they should be of approximately the same magnitude. Therefore, in view of (61), an improvement of the ω-resolution by a certain factor m requires a reduction of δE_0 *and* δE_1 by m. Consequently, the counting rate Z (or the intensity) at the detector is reduced by m^2. Similar arguments lead to the conclusion that an improvement of the κ-resolution by a factor m leads to a reduction of the intensity by m^2 for isotropic systems. In unfavourable cases (for single crystal work) the reduction is even m^3 or m^4. As a consequence, if the energy resolution is adapted[5] to the resolution in κ, the count rate at the detector is at least proportional to the 4th power of the overall resolution width.

---

[5] For instance, in many problems of quasielastic scattering the width Γ_{exp} of the quasielastic line is proportional to κ^2. Therefore, $\delta\kappa/\kappa$ should be made equal to $\delta\Gamma_{exp}/2\Gamma_{exp}$, in order to achieve an overall optimum layout of the experiment. Practically, in most cases one would choose only a fraction of this value.

The relative statistical error $\sqrt{\langle \delta Z^2 \rangle}/Z$ of a count rate is proportional to the inverse square root of the total number of counts, ZT_0, where T_0 is the available experimental time. In particular one can show that the statistical uncertainty with which a line width Γ_{exp} (including resolution broadening) can be determined, is [27]

$$\delta \Gamma_{\text{exp}}/\Gamma_{\text{exp}} \simeq 0.7 \left(\sum_{v=1}^{n} Z_v T_0 \right)^{-1/2} \tag{62}$$

assuming that the spectrum has been determined with the help of n aequidistant values Z_v of the count rate, each measured over the same time interval T_0.

Assume now that a certain statistical accuracy, i.e. a certain statistical error, is demanded. Then an improvement in the overall resolution by a factor of say $m = 3$ requires an increase of the experimental time T_0 by at least 3^4 for a given strength of the neutron source! This demonstrates that even a modest improvement of the resolution tremendously increases the time required for the experiment.

The possibilities of improving the *sources* are restricted by technological problems and by costs. The fluxes of conventional research reactors are of the order of a few 10^{13} to $10^{14}\ \text{sec}^{-1}\ \text{cm}^{-2}$. The highest values which can be achieved in modern high-flux reactors are about $10^{15}\ \text{sec}^{-1}\ \text{cm}^{-2}$. On the other hand, one has to keep in mind that relevant improvements are also feasible with respect to the instrumentation. By using, for instance, multidetectors, focusing methods, and better crystals one is able to gain another order of magnitude in experimental time.

3.2 Instruments

The monochromatization of the incident beam and the analysis of the scattered neutrons can be performed by means of Bragg reflections on single crystals. Alternatively, this can be achieved by chopping the incident beam into short pulses and by performing a time-of-flight analysis of the scattered neutrons. There are instruments using Bragg reflections both before and after scattering (triple axis spectrometers) or time-of-flight in both cases (double choppers), and combinations of both methods (see [3, 4]). Here we will restrict ourselves to the description of two more recent developments which are of special importance for measurements at small energy transfers.

"*90° spectrometer*". The wave-number k of a neutron beam diffracted on a single crystal with a Bragg angle θ is given by Bragg's law $G = 2k\sin\theta$, where G is a reciprocal lattice vector ($G = 2\pi/d_{hkl}$). The mean-square energy width of the reflected beam is found from the derivative of this equation, namely

$$(\langle \delta E^2 \rangle)^{1/2}/2E = (\text{ctg}^2 \theta \langle \delta \theta^2 \rangle + \langle \delta G^2 \rangle/G^2)^{1/2} . \qquad (63)$$

The square angular width of the Bragg angle, $\langle \delta \theta^2 \rangle$, is due to the mosaic spread of the crystal and the beam divergency. The term $\langle \delta G^2 \rangle$ originates from the fact that only a finite number of lattice planes is contributing to a Bragg reflection. This leads to a broadening of the reciprocal lattice points. Normally, high resolution is being achieved by reducing the divergency of the beam which, correspondingly, reduces the intensity.

The other way is to choose $\theta \simeq \pi/2$ as Bragg angle. Consequently, $\delta \theta$ can be relatively large, and the resolution is only determined by $\delta G/G$ which is of the order of 10^{-4}. Due to the fact that θ is fixed, the energy has to be varied by means of a Doppler motion of the monochromator or the analysing crystal. According to a proposal of Maier-Leibnitz, a neutron spectrometer has been constructed on this basis at the FRM reactor in Munich. Its resolution (full width at half maximum) is about $\delta \hbar \omega \simeq 7 \cdot 10^{-7}$ eV, and the energy range to be covered by the Doppler motion is about $\pm 2 \cdot 10^{-6}$ eV [28]. A similar and improved version of such a spectrometer with a resolution of $4 \cdot 10^{-7}$ eV is shown in *Fig. 7*. With this kind of instrument one obtains a broad overlap of the region of neutron spectrometers and that of relaxation methods (Fig. 6).

Statistical chopper. In conventional time-of-flight spectrometers, an incident monochromatic neutron beam is chopped periodically into short pulses; their duration is of the order of 10^{-4} to 10^{-5} sec, with repetition rates of the order of 10^2 per second. The intensity distribution $Z(t)$ is then measured as a function of flight time t of the scattered neutrons between sample and detector *(Fig. 8)*. Neglecting frame-overlap due to neutrons arriving from bursts of previous cycles, one gets a convolution

$$Z(t) = \mathscr{A} \int_0^\infty \mathscr{S}(t') R(t - t') \, dt' + b(t) . \qquad (64)$$

$\mathscr{A}$ is some constant of proportionality. $b(t)$ is the background. The function $\mathscr{S}(t')$ is the scattering probability. It is proportional to the scattering cross section $d^2\sigma[E_0, E_1(t'), \vartheta]/d\Omega dE_1$ to be measured; E_1 is $mL^2/2t'^2$ and dE_1 is proportional to $\delta t'/t'^3$; $\delta t'$ is the channel width of the flight-time analyzer. The function $R(t)$ is the intensity of the neutron beam as a function of time during the pulse. The width of this function, $\Delta \tau_P$, determines the instrumental *resolution*. Obviously, $Z(t)$ approaches $\mathscr{S}(t)$ for very small $\Delta \tau_P$. If one scatters on a solid $Z(t)$ equals $R(t)$ because the energy spectrum contains a delta-function at $\hbar \omega = 0$. To avoid too much frameoverlap, the distance τ_R of successive cycles must be suf-

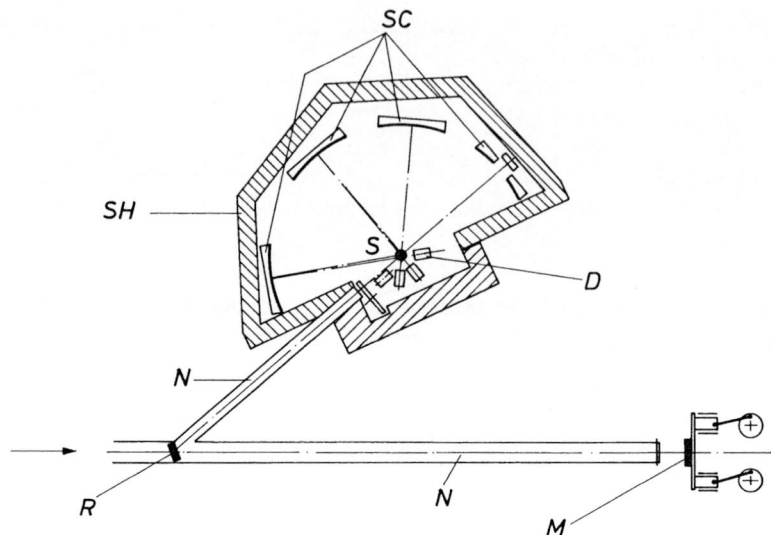

Fig. 7. "90°-spectrometer" (at FRJ-2 Dido-reactor). *R* reflector crystal; *N* neutron con-
ducting guides; *M* monochromator crystal with piston motor for producing a Doppler
shift of energy E_0; *S* sample; *SH* shielding; *SC* arrangement of selector crystals; *D* detec-
tors. Bragg angle of selector and analyzer approximately 90° (From Alefeld [29])

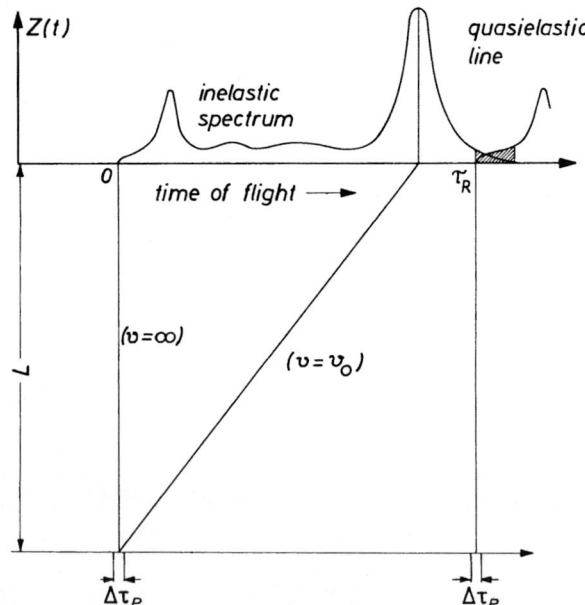

Fig. 8. Diagram for a time-of-flight spectrometer. $\Delta\tau_P$ pulse length; τ_R pulse repetition
interval; $v = v_0$ corresponds to elastic scattering. $Z(t)$ time-of-flight spectrum. Shaded:
Region of frame-overlap

ficiently large. This means, that only a small fraction $\Delta\tau_P/\tau_R$ of the incident intensity is being used. In typical cases $\Delta\tau_P/\tau_R$ is of the order of 10^{-2}.

The "time utilization" can be considerably improved by the method of statistical chopping [30–33] where we replace $R(t)$ in (64) by a non-periodic modulation $M(t)$; it can be characterized by its autocorrelation function

$$\mathscr{C}(\tau) = \lim_{t_0 \to \infty} (1/2t_0) \int_{-t_0}^{t_0} M(t')\, M(t' + \tau)\, dt' . \tag{65}$$

If one now cross-correlates the measured time-dependent $Z(t)$ at the detector with the modulation function $M(t)$, one finds a function $K(\tau)$ which is according to (64) and (65)

$$K(\tau) = \lim_{t_0 \to \infty} (1/2t_0) \int_{-t_0}^{t_0} Z(t')\, M(t' - \tau)\, dt' \equiv \int_{0}^{\infty} \mathscr{S}(t')\, \mathscr{C}(t - t')\, dt + b\langle M \rangle . \tag{66}$$

Here $\langle M \rangle$ is the mean value of the modulation $M(t)$ and b is assumed to be a constant. The cross-correlation of $Z(t)$ with $M(t)$ can be performed on-line by electronical means. The resulting function $K(\tau)$ is a convolution of the scattering probability with the auto-correlation function $\mathscr{C}(\tau)$ which is a known property of the instrument.

If a random function $M(t)$ according to "white noise" could be realized (where $\mathscr{C}(\tau) = \mu\delta(\tau) + \text{const}$), the function $K(\tau)$ in Eq. (66) would directly reproduce the spectral function $\mathscr{S}$. Such a random modulation can not be generated because of the finite switching time of any neutron chopping device. It is reasonable to approximate this auto-correlation function by modulating the beam according to a *pseudorandom sequence* of bursts. These are of finite width and can be realized by alternating transparent and adsorbing sections on a rotating disc [33].

The important advantage of a statistical chopper device follows from the fact that the opening ratio, and, therefore, the intensity at the detector can be improved by at least one order of magnitude above its value $\Delta\tau_P/\tau_R \simeq 10^{-2}$ for a periodical chopper. A "gain factor" can be calculated as defined by the ratio of the mean square statistical error of the spectral function $\mathscr{S}(\tau)$ for the periodical and for the statistical chopper, namely

$$g^2 = \langle \delta\mathscr{S}(\tau)^2 \rangle_{\text{period}} / \langle \delta\mathscr{S}(\tau)^2 \rangle_{\text{statist}} . \tag{67}$$

This is shown in *Fig. 9* as a function of the opening ratio and of the signal height per time interval. The result depends strongly on the level of the uncorrelated background, i.e. the background *not* due to scattering of the incident neutrons on the sample.

For a quasielastic scattering peak, the ratio g^2 is particularly high, especially for small κ where most of the scattering is concentrated around

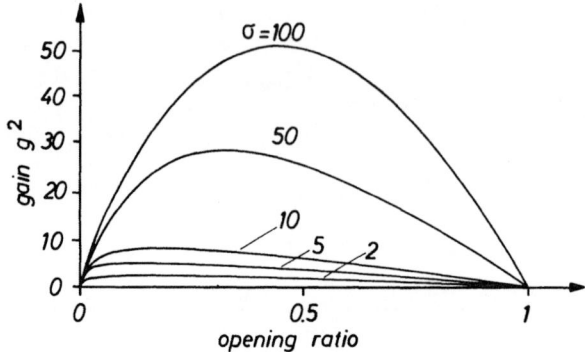

Fig. 9. Gain factor with respect to statistical error of the signal height $\mathscr{S}(\tau)$, for a statistical chopper compared to a periodical chopper; $g^2 = \langle \delta \mathscr{S}(\tau)^2 \rangle_{\text{period}} / \langle \delta \mathscr{S}(\tau)^2 \rangle_{\text{statist}}$ as a function of the opening reatio. σ is the relative signal height; $\sigma = \mathscr{S}/\langle \mathscr{S} \rangle$. The curves are calculated for a relative background $b/\langle \mathscr{S} \rangle = 1$ (From Hoßfeld [33])

$E \approx E_0$. In favourable cases, values for g^2 up to ten might be achieved for opening ratios of the order of 0,5. The technical and theoretical development of statistical chopping devices is still in an early stage; further progress in the application of this method is to be expected.

3.3 Corrections

The intensity measured in neutron scattering experiments does not directly reproduce the cross-section (15). Several corrections and other mathematical manipulations have yet to be applied. The most important of them will be discussed briefly.

(i) Resolution corrections. These have already been formulated for time-of-flight spectrometers by the convolution integrals (64) and (66). For triple-axis spectrometers we refer to [34]. If only the half-width of a Lorentzian is to be determined, and if it can be assumed that the resolution function is Gaussian-shaped, the method given by Melkonian [35] can be applied. As will be seen later, in many cases the spectrum is more involved, sometimes being a sum of Lorentzians of various widths; then, a theoretically predicted or guessed function can be used where the parameters are to be determined by a least-squares fit. Alternatively, the measured spectrum has to be unfolded. The latter method (see for instance [36]) is mathematically very complex, in particular with respect to error propagation. Resolution corrections with respect to κ are often negligible due to the smooth κ-dependence of the scattering law. Also the dependence of κ on E_1 (for fixed scattering angle ϑ) can often be neglected (see Eq. (12)).

(ii) Multiple scattering There exists a certain probability that neutrons are scattered in the sample more than once. For double scattering, which is the dominating process, the resulting intensity distribution is then a self-convolution of the scattering law. The resulting double-scattering spectrum can be calculated to first order from the measured uncorrected single scattering spectrum (see for instance [37]). An estimate of the magnitude of multiple scattering (without regarding its energy distribution) can be found in [38]. Very roughly, the double scattering contribution can easily be as high as 30% of the single scattering intensity, if the latter is, in a typical case, 10% of the primary flux.

(iii) Background. Background corrections are not trivial. In particular, the presence of the sample might influence the strength of the background so that a "sample-out" experiment is not necessarily representative for the correction to be applied. A further difficulty is the appearance of elastic Debye-Scherrer peaks from the sample holder. Background corrections are especially important for the wings of the quasielastic spectra.

(iv) Absolute intensity scale. In many problems it is of extreme importance for the relevance of the data to get the scattering cross section in an absolute scale, for instance in barn/steradian, and not in arbitrary units. This can be achieved by the use of "standard scatters" (for instance [37]) like vanadium which has a well-known scattering cross section; for time-of-flight spectrometers a calibration is feasible by calculating the detector efficiency, chopper transmission, and air absorption.

Resolution and multiple scattering corrections can always be reduced on the expense of intensity, and therefore, on the expense of statistical accuracy. Consequently, in the preparation of an experiment one has to achieve an optimum compromise between statistical accuracy on the one hand, and reliability of the results with respect to systematical error sources on the other.

3.4 Separation of S_{coh} and S_{inc}

Table 1 represents examples of bound scattering cross-sections σ_{coh} and σ_{inc} according to the definitions (14) with (13) and (17). Hydrogen, due to the strong spindependence of the neutron-proton interaction, has an incoherent cross-section which is extraordinarily large as compared to the coherent cross-section; it is also large compared to σ_s of practically all other elements. Therefore, scattering on hydrogeneous compounds gives, to a good approximation, $S_{inc}(\kappa, \omega)$ which yields information concerning the *individual* motion of particles. The large hydrogen scattering cross-section also allows the investigation of the motion of certain atomic groups in a molecule if these groups are "labelled" by hydrogen atoms, or if all hydrogens outside these groups are replaced by atoms with small cross-sections. This *substitutional technique* has been extensively applied by White and others [39, 40].

A problem of particular interest is to investigate separately S_{inc} and S_{coh} at the same substance. According to (15), $d^2\sigma/d\Omega dE_1$ is a linear

Table 1. Important bound cross-sections in barns (1 barn $= 10^{-24}$ cm^2)[a]

Element	σ_{coh}	σ_{inc}
H	1.79	79.7
D	5.4	2.2
C	5.50	[b]
O	4.2	[b]
Na	1.55	1.85
Ar[c]	0.5	0.4

[a] For complete compilations see [41, 2].
[b] Negligibly small.
[c] For Ar^{36} see [42].

combination of both functions. By changing the isotope composition, a separation is possible if the scattering lengths of the isotopes differ sufficiently. Such experiments have been performed for the first time on liquid argon [43]. For monoisotopic nuclei with nonzero spin and with sufficiently strong spin incoherence (for instance H, D, Li^7, As, Na, V) a separation is, in principle, also possible due to the following fact: Spin incoherent scattering occurs, in contrast to coherent scattering, with spin reversal [1]. Therefore, a separation of S_{inc} and S_{coh} is possible by means of a spectrometer with a spin polarizer and a spin analyzer system. As a matter of fact, such experiments need very intense neutron sources.

4. Monoatomic Liquids with Continuous Diffusion

During the last decade, the study of simple liquids by quasielastic and inelastic scattering was one of the most important activities in neutron spectroscopy (see [3, 4, 44, 45]). In addition to results already existing from the classical relaxation methods, qualitatively new information has been obtained concerning the *short-time* behaviour of the atomic motions. On the present level of experimental accuracy, it seems that a certain saturation has been reached in this subject, and existing theories and experiments will be reviewed here and in Section 5. More complicated liquids, where rotational degrees of freedom play an important role, will be discussed later in Section 8 and 9.

4.1 The Langevin Equation

First we consider atoms of mass M in an ideal gas with temperature T, performing a free-flight motion. The self-correlation function is then rigorously Gaussian-shaped and one obtains from (39) a time-dependent width function

$$\gamma(t) = \langle v^2 \rangle \, t^2/6 = (k_B T/2M) \, t^2 \, . \tag{68}$$

If there exists a mutual interaction between the particles, each of them is coupled to its neighbours. The calculation of the correlation functions would correspondingly require the solution of the equations of motion for all coupled atoms. To simplify the problem, one can try to replace the interaction with the neighbours of a certain atom by a rapidly fluctuating effective force. According to the classical Langevin theory of Brownian motion [21, 46] this force should fulfill the following requirements: If averaged over an ensemble of many particles with a given velocity $v(t)$, it should vary slowly with time; furthermore, it should be proportional to $-v(t)$ according to the friction force in Stokes' law. The rapidly fluctuating component of the force, which we call $F(t)$, is cancelled in the ensemble average. This force is supposed to be uncorrelated with the velocity of the particle under consideration. Formally this means that [47]

$$\langle F(t) \rangle = 0, \quad \langle F(0) \, v(t) \rangle = 0, \quad \text{and} \quad \langle F(0) \, F(t) \rangle = 0, \tag{69}$$

except for very small t. Therefore, the total force acting on the particle under observation can be written as $-(v/B) + F(t)$ and the equation of motion reads then

$$M \, dv/dt = -v/B + F(t) \, .$$

With $\eta = 1/MB$ and $A = F/M$ one gets

$$dv/dt = -\eta v + A(t) \, . \tag{70}$$

Integration gives directly

$$v(t) = v(0) \exp\{-\eta t\} + \exp\{-\eta t\} \int_0^t \exp\{\eta t'\} \, A(t') \, dt' \tag{71}$$

for an initial velocity $v(0)$. The velocity autocorrelation function is found by multiplying this equation with $v(0)$ and averaging it subsequently, whereupon one gets for $t > 0$

$$\langle v(0)\, v(t)\rangle = \langle v^2\rangle \exp\{-\eta t\} = (3k_B T/M)\exp\{-t/\tau_r\}\,. \tag{72}$$

Here we have introduced

$$\tau_r = 1/\eta\,, \tag{73}$$

the relaxation time of the velocity correlation. For $t \gg \tau_r$, the particle has lost the memory of its initial velocity vector and consequently $\langle v(0)\, v(t)\rangle$ approaches zero.

With (39) and (72) one finds the width function

$$\gamma(t) = D[t - \tau_r(1 - \exp\{-t/\tau_r\})] \tag{74}$$

where we have introduced the macroscopic self-diffusion constant[6]

$$D = k_B T/\eta M = \tau_r k_B T/M\,. \tag{75}$$

For $t \ll \tau_r$, Eq. (74) approaches $\gamma = (k_B T/2M)\, t^2$ as for an ideal gas. For $t \gg \tau_r$ one gets

$$\gamma(t) = D(t - \tau_r)\,. \tag{76}$$

Therefore, τ_r is a kind of delay before the average behaviour of the particles becomes purely diffusive. Neglecting the term τ_r, one recognizes that $G_s(r, t)$ according to (30) and (76) is a solution of the diffusion equation

$$D \Delta G_s(r, t) = (\partial/\partial t)\, G_s(r, t)\,. \tag{77}$$

From (22) the corresponding scattering law is found to be a Lorentzian-shaped quasielastic line

$$S_{\text{inc}}(\kappa, \omega) = \int\limits_{-\infty}^{\infty} \exp\{i\omega t - \kappa^2 \gamma(t)\}\, dt = \frac{\kappa^2 D/\pi}{\omega^2 + (\kappa^2 D)^2}\,. \tag{78}$$

Its area is unity, and its half-width at full-maximum is

$$\Gamma = 2\hbar\kappa^2 D\,. \tag{79}$$

This simplified description of scattering on diffusing atoms has been derived for the first time by Vineyard [16].

---

[6] $F(t)$ enters (72) only implicitly: A necessary requirement is that v should be distributed for large t like a Maxwellian with temperature T. This leads, by means of (71), to a connection between the random force F and the friction constant [46], namely $\int\limits_{0}^{\infty} A(0)\, A(t')\, dt = 2D\eta^2$.

4.2 Oscillatory Diffusion

Primarily, the Langevin Eq. (70) is applicable to the Brownian motion of a particle embedded in a gas where the time between the impacts the particle suffers is short compared to the observation time[7]. Now we consider an *atom* in a liquid, i.e. the case of a particle being identical with the surrounding particles, and we want to study the velocity correlations for times below 10^{-11} sec (for a survey on the various theoretical concepts see e.g. [47]). In this situation, the significant improvement to be introduced is an *enlargement of the decay time* of the function $\langle v(0) \, v(t) \rangle$. Physically this delay is due to a strong correlation of the atom with its closest neighbours. In general, this correlation is also related to an *oscillatory component* of the motion. These considerations are quantitatively taken into account in the models to be discussed subsequently.

The inadequacy of the description by the Langevin model can be inferred quantitatively from the velocity correlation time τ_r as found from (75). For a typical liquid $(D \simeq 2 \cdot 10^{-5}\,\mathrm{cm^2/sec})$ τ_r would be about 10^{-13} sec. This is the same order of magnitude as the oscillation periods of the corresponding solid. In reality, a particle with a certain excess velocity could never reach its average velocity in such a short time; it would rather perform oscillations in the field of its neighbours until its excess velocity has decayed.

The failure of the simple Langevin theory has also been verified by early quasielastic neutron scattering experiments on liquids (for instance [48, 50, 51]). As will be discussed later, it has been found that, for small κ, the quasielastic width $\Gamma(\kappa)$ agrees with (79), whereas for large κ it is smaller than predicted by the Langevin theory. Furthermore, the existence of oscillatory components of motion has clearly been demonstrated by the occurence of peaks in the inelastic neutron spectra, similar to those appearing in a solid.

In view of the arguments given before a simple heuristic model has been proposed by Egelstaff and Schofield [52, 53]. The width function $\gamma(t)$ is separated into a *diffusive* and an *oscillatory component*, namely

$$\gamma(t) = \gamma_{\mathrm{D}}(t) + \gamma_{\mathrm{osc}}(t) . \tag{80}$$

The latter can be approximated by

$$\gamma_{\mathrm{osc}}(t) = \gamma_{\infty} [1 - (\sin 2\pi v_{\mathrm{D}}t)/(2\pi v_{\mathrm{D}}t)] \tag{81}$$

as for a harmonic Debye crystal with a characteristic frequency v_{D} [16]. The quantity

$$\gamma_{\infty} = \langle u^2 \rangle = 3k_{\mathrm{B}}T/M(2\pi v_{\mathrm{D}})^2 \tag{82}$$

is the *mean square amplitude* of the vibrating atom (or the mean square radius of the "thermal cloud" which is built up by this motion).

With respect to $\gamma_{\mathrm{D}}(t)$ the following picture is being used. The motion of a diffusing atom in liquids is strongly correlated with the motion of its

---

[7] To be more quantitative, we may state: The velocity auto-correlation found from (70) is correct for times large compared to the time intervals during which the force correlations (69) have not yet decayed.

neighbours for times $t > 1/\nu_D$. This correlation is, in a rough approximation, described by the Brownian motion of a rigid *cluster* or "globule" of N_C atoms; therefore, in the Langevin equation an enhanced mass $\hat{M} = M N_C$ it is to be used.

As a consequence, the relaxation time τ_r in (75) is increased by a factor N_C, being now

$$\hat{\tau}_r = 1/\hat{\eta} = (MD/k_B T)(\hat{M}/M) > 1/\eta .\tag{83}$$

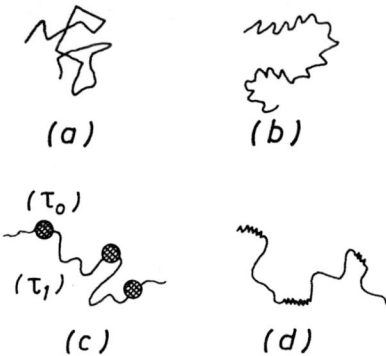

(a)   *(b)*

(τ_0)

(τ_1)

(c)   *(d)*

Fig. 10a–d. Sketch of particle paths for various kinds of diffusion. a Brownian motion according to the Langevin equation. b Superposition of diffusive and oscillatory motion. c Step-wise motion with oscillatory and diffusive periods. Shaded circles symbolize the "thermal cloud" which has developed by the vibrational motion during τ_0. d Itinerant oscillator.

The radius of such a cluster should be considered as the average correlation distance for the particle motion. As a matter of fact, this correlation exists only for a limited period of time. The diffusive part of $\gamma(t)$ is then formulated as

$$\gamma(t) = D[(y^2 + \hat{\eta}^{-2})^{1/2} - \hat{\eta}^{-1}]\tag{84}$$

Here, $y = [t^2 - (i\hbar t/k_B T)]^{1/2}$ is inserted instead of t so that $G_s(r, t)$ has the correct quantum mechanical behaviour for short times (see Section 2.2). The particular mathematical form (84) rather than (74) has been chosen in order to simplify the mathematical treatment; it is used very frequently in theoretical work. Eq. (84) gives the proper behaviour for large and small t, otherwise it does not deviate strongly from the Langevin formula (74).

In contrast to formula (75), $\hat{\eta}$ is now a *disposable parameter* so that it is connected with D by (83) rather than (75).

Fig. 10 illustrates, in an artistic way, the paths of atoms for the different kinds of motion to be considered here and in Section 5. *Fig. 11* sketches

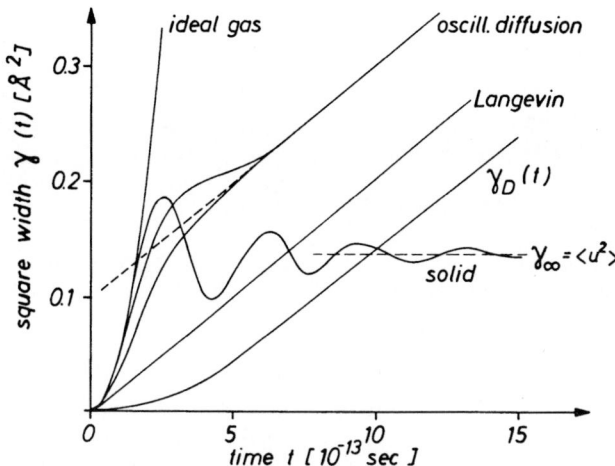

Fig. 11. The square width $\gamma(t) = \langle r^2(t)\rangle/6$ of a Gaussian self-correlation function (Eq. (30)) for the ideal gas, the Langevin motion, the solid Debye crystal, and the oscillatory diffusion model of Rahman et al. [54] (for two different choices of the damping parameter). $\gamma_D(t)$ is the diffusive part of $\gamma(t)$ for the effective mass model [52]. The drawing is schematic, but the scale gives the correct order of magnitude for a normal liquid

the width functions $\gamma(t)$ for the ideal gas, the Langevin and the cluster model. The scattering law $S_{\text{inc}}(\kappa, \omega)$ for the cluster model is, for small κ, a Lorentzian as in (78), with $\Gamma = 2\hbar\kappa^2 D$. For large κ it approaches a Gaussian with a width at half maximum of

$$\Gamma = 2\hbar(2\ln 2)^{1/2} \, (\kappa^2 D/\hat{\tau}_{r1})^{1/2} \tag{85}$$

$\hat{\tau}_{r1}$ is a modified delay time given by

$$\hat{\tau}_{r1}^2 = \hat{\eta}^{-2} + (\hbar/2k_B T)^2 \,. \tag{86}$$

Consequently, an enhancement of the delay time $\hat{\tau}_{r1}$ is able to reduce the half width as suggested by the experimental results. In this model the behaviour of the scattering law and the width show the same qualitative features as those of the spectra in Fig. 17.

A similar quasi-crystalline model of oscillatory diffusion has been worked out by Rahman et al. [54]. The position of the particle $r(t)$ is described by a linear superposition of statistically independent components $\xi_\nu(t)$ which should obey equations of motion with restoring forces $-\omega_\nu^2 \xi_\nu(t)$ and stochastic driving forces $F_\nu(t)$, namely

$$d^2\xi_\nu/dt^2 + \eta_\nu d\xi_\nu/dt + \omega_\nu^2 \xi_\nu = F_\nu(t) \,. \tag{87}$$

Such an equation does not lead to any diffusive motion. Therefore, it is necesseary to introduce, for a certain fraction of the degrees of freedom, a motion without restoring force as in (70), namely

$$d^2\xi_\nu/dt^2 + \beta_\nu d\xi_\nu/dt = F_\nu(t) \,. \tag{88}$$

In a liquid, transversal sound waves might exist at *high* ω; therefore it is plausible to assume that Eq. (87) holds for ω_v above a certain frequency ω', whereas (88) holds for $\omega_v < \omega'$. Here, ω' is a disposable parameter of the model (for simplification one has not taken into account that longitudinal modes exist also for small ω_v). Solving these equations [46] and summing up all modes in both regions with a weight function according to a Debye spectrum $f_D(\omega)$ one obtains

$$\langle v(0)\, v(t)\rangle = 3 \int_0^{\omega_D} f_D(\omega_v)\langle \dot{\xi}_v(0)\, \dot{\xi}_v(t)\rangle\, d\omega_v. \tag{89}$$

From this equation, the width function $\gamma(t)$ can be calculated by means of (39). Its qualitative features are shown in Fig. 11. For the asymptotic form of $\gamma(t)$ one has

$$\gamma(t) = D(t - \eta^{-1}) + \gamma_\infty [1 - (\omega'/\omega_D)] \tag{90}$$

where

$$D = (k_B T/\hat{\eta} M)\,(\omega'/\omega_D)^3. \tag{91}$$

γ_∞ is the mean square vibrational amplitude of the atoms in a Debye-crystal if all normal modes are of the oscillatory type.

For an experimentally given diffusion constant D, the delay time $1/\hat{\eta}$ can be enlarged by allowing that a certain fraction of the modes, $1 - (\omega'/\omega_D)^3$, is of the oscillatory type; the other modes are diffusive. In this model, Eq. (75) is modified by a factor $(\omega'/\omega_D)^3$ in analogy to the factor $(M/\hat{M})$ of the Egelstaff-Schofield model. $M/\hat{M} = (\omega'/\omega_D)^3 = 1$ yields the Langevin case.

Finally we quote the *itinerant oscillator model*, as developed and worked out by Sears [55] and others [56, 57]. It introduces the concept of a center of oscillation, $R(t)$, around which damped oscillations occur. $R(t)$ is assumed to undergo a slow diffusive motion as described by an equation of the Langevin type (70); the effective frequency Ω_0 of the oscillatory component is calculated according to (57).

A separation of the width function into some kind of diffusive and into an oscillatory part as in (80) will be used throughout all theories to be dealt with here. This means that both kinds of motion are considered as *uncorrelated*. As a consequence, one obtains a factorization of the intermediate scattering function [52, 37], namely

$$I_s(\kappa, t) = \exp\{-\kappa^2 \gamma_D(t)\}\, \exp\{-\kappa^2 \gamma_{osc}(t)\} \tag{92}$$

and, correspondingly, a convolution for the scattering law which reads

$$S_{inc}(\kappa, \omega) = \int_{-\infty}^{\infty} S_{inc}^D(\kappa, \omega')\, S_{inc}^{osc}(\kappa, \omega - \omega')\, d\omega' \tag{93}$$

$S^D(\kappa, \omega)$ is the Fourier transform of the diffusive part which is the bell-shaped quasielastic line. The factor $\exp\{-\kappa^2 \gamma_{osc}(t)\}$ can be expanded in powers of $\kappa^2 \gamma_\infty$ which corresponds to a *phonon expansion* as it has been developed for harmonic crystals [58]

$$S_{inc}^{osc}(\kappa, \omega) - \exp\{-\kappa^2 \langle u^2\rangle\}\, \{\delta(\omega) + \kappa^2 (2\pi)^{-1} \int [\gamma_\infty - \gamma(t)]\, \exp\{-i\omega t\}\, dt$$
$$+ \kappa^4 [...]\}. \tag{94}$$

Here, $\delta(\omega)$ describes the "zero phonon" term where only momentum, but no energy is exchanged; the term proportional to κ^2 is due to "one-phonon" transitions, and so on for further terms. For a harmonic crystal with a frequency spectrum $f(\omega)$, the one-phonon terms can be written as

$$(\kappa^2 k_B T / 2 M \omega^2) f(\omega).$$

After having performed the convolution (93), the $\delta(\omega)$-term reproduces the quasielastic line with an intensity proportional to a Debye-Waller factor, $\exp\{-\kappa^2 \gamma_\infty\} \equiv \exp\{-\kappa^2 \langle u^2 \rangle\}$. This quasielastic line rides on the broad distribution of one- and multi-phonon transitions which is "flat", i.e. ω-independent for small ω. For liquids which lack in pronounced oscillatory or quasi-crystalline behaviour, the quasielastic line can not be separated from the inelastic "background", especially not for large κ-values (see Fig. 17).

4.3 The Memory Function Concept

In an earlier section the diffusive motions of atoms in a liquid have been described by the Langevin equation; subsequently, this equation has been improved by the introduction of restoring forces which have led to some oscillatory motion, and to a delay of diffusion. In the following, we briefly present a generalization of the Langevin equation which is able to reveal these effects on a more general basis. This concept has been developed and applied by a great number of authors [59–64, 19, 47]; it is expected to be helpful for the interpretation of scattering experiments.

It has been derived that the velocity $v(t)$ of a certain particle in a many-particle system, and its auto-correlation function $\psi(t)$ obey the following rigorous equations [59, 19]

$$d v(t)/d t = -\int_0^t K(t - t') v(t') d t' + F(t)/M \tag{95}$$

and

$$d \psi(t)/d t = -\int_0^t K(t') \psi(t - t') d t'. \tag{96}$$

Analogous formulations hold also for orientational correlations [19]. $F(t)$ and $K(t)$ are functions which, in principle, could be calculated from the interaction potential between all particles of the system.

The kernel $K(t)$, which characterizes the dynamical behaviour of the atoms in the system, is a "memory function": According to (95) or (96), the decay rate of the velocity, or of $\psi(t)$, at a given instant t, is determined by the value of v or ψ taken at previous times $t' < t$, and not only at instant t, depending on the time interval during which $K(t - t')$ has not yet decayed. This behaviour can be qualitatively understood in the following sense. During its motion a particle leaves behind a disturbancy of its neighbourhood; the reaction on the particle resulting from this disturbancy is active within a *finite* time interval after its creation. Obviously, if the "memory" is lost instantaneously, so that $K(t - t') = \delta(t - t')$, one obtains the familiar Langevin equation (70) with (72). The integral $\int_0^\infty K(t')\,dt'$ has then to be chosen as the friction constant.

One can construct memory functions from simple phenomenological expressions [61, 65]; it has turned out, for instance, that the mere introduction of a finite memory time τ_m in a decay function

$$K(t) = \gamma_0 \exp\{-|t|/\tau_m\} \tag{97}$$

is already able to reproduce an oscillatory behaviour of $\psi(t)$[8]. On the other hand, it can also be tried to compute $K(t)$ approximately on a more first-principle basis [66, 67], for instance from the interatomic potential and the pair correlation function. In general, it turns out that K has a rapid initial decay within a few 10^{-13} sec which is supposed to be determined by the time of the hard-core collisions, followed by a slow and monotonic decay as a consequence of soft and long range interactions.

4.4 Computer Experiments

High speed computers with large memories have opened the possibility to perform mathematical "experiments" on liquids and on other many-body systems. By solving the classical equations of motion for a finite, but relatively large number of particles one obtains the static equilibrium quantities of this system, as well as transport coefficients and time-dependent correlation functions. Such "experiments" have been widely used as tests for model theories, in particular because the interactions are known *a priori;* in many respects, these experiments are able to serve as a guide line for new theoretical models. Very often, they have given deeper insight into the arrangement and the motion of atoms in liquids than real experiments, because of the insufficient accuracy of the latter.

For the first time, computer experiments on the static and dynamic behaviour of liquid noble gases have been carried out by Rahman [14, 68] and by Verlet et al. [69, 70] using a simple van der Waals-interaction as applicable for noble gas atoms. Computations concerning liquid metals [71, 72] have been performed approximating the interaction by a two-body pseudo-potential; it includes a long-range oscillatory contribution like $r^{-3}\sin(2k_F r)$, where k_F is the Fermi wave-number. In the following, we restrict ourselves to a brief description of the nature of this method; its application will be mentioned several times in the subsequent sections.

---

[8] τ_m is not to be confused with the delay time τ_r in (75).

The time-dependent position $r_i(t)$ of a particle with a mass M is described by Newton's equation

$$M \, d^2 r_i / d t^2 = - \sum_{j=1}^{N} \partial V(|r_i - r_j|)/\partial r_i \tag{98}$$

where V is the interaction potential between the particles.

A certain number N of particles is enclosed in a cubic box of edge length L so that N/L^3 determines the density of the liquid. By applying periodic boundary conditions one gets an infinite system of equations, but only the equations for N particles have to be solved. By truncating $V(r_{ij})$ somewhere, the sum in (98) goes only over a finite number of neighbours. A sufficiently large number of particles has to be chosen to obtain a good statistical accuracy (about 10^3 in Rahman's calculations). The initial particle velocities $v_i(0)$ are selected according to a Maxwell distribution in such a way that the temperature of the system has the required value.

After a sufficiently long time has elapsed, the positions and velocities of all N particles at successive time intervals are collected; thus the quantities of interest can be computed. For instance, the time-dependent pair correlation function $G_d(r, t)$ can be determined according to its (classical) definition (28b) for an isotropic system,

$$G_d(r, t) \, 4\pi r^2 \Delta r = \langle n(r, t) \rangle . \tag{99}$$

Here $\langle n(r, t) \rangle$ is the average number of particles, as situated for a time t at a distance between r and $r + \Delta r$ from the position which was occupied by a certain atom at $t = 0$. Furthermore, the velocity auto-correlation function $\langle v(0) \, v(t) \rangle$ can be evaluated; the average can be taken both on the N particles and on the time origin. From the calculable moments in space $\langle [r(t) - r(0)]^{2n} \rangle$, the self-correlation function $G_s(r, t)$ can be reconstructed [69]. It turns out that computer experiments gave fairly good agreement with experimental data, for instance with the self-diffusion constant and their temperature dependence; a comparison of $\langle v(0) \, v(t) \rangle$ with neutron scattering experiments has also been performed (see Section 4.5).

Computer experiments can be extended to more complicated systems, like simple molecular liquids, by introducing a rotational degree of freedom, and by including angular dependent multipole potentials [73]. Such calculations are able to reveal also the time-dependent orientational correlation functions of the molecular axes (Section 7.2). In principle, it might also be feasible to introduce many-body forces which are, by no means, negligible.

Severe limitations exist with regard to the behaviour of the atoms at large times and distances because of the limited capacity of computers;

special attention has been paid to such investigations by means of computer experiments in systems with simplified interactions. [74]. The investigation of quantum liquids seems to be far beyond the capacity of computers and of existing numerical methods.

4.5 Experiments on Liquid Sodium and Argon

Liquid sodium has extensively been investigated by Cocking [37] and by Randolph [25]. *Fig. 12* compares typical time-of-flight spectra for the solid and the liquid phase. For the solid one recognizes the elastic line;

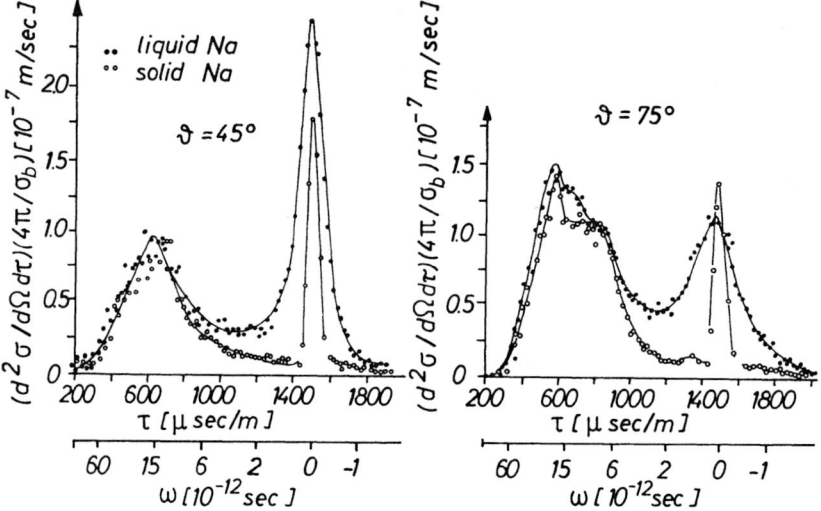

Fig. 12. Typical scattering cross section per time-of-flight interval $d^2\sigma[E_0, E_1(\tau), \vartheta]/d\Omega\,d\tau$ in units of the bound cross sections σ_b as a function of flight time τ per meter and of energy transfer $\hbar\omega$, for sodium as a solid (open circles) and a liquid (full circles). Intense peak at $\hbar\omega = 0$ is the elastic or quasielastic line. The width of the elastic line in the solid is given by the instrumental resolution. The elastic lines for the solid have been reduced in ordinate by a factor 4; $\vartheta =$ scattering angle (From Cocking [37])

its finite width is determined by the instrumental resolution. In the liquid phase this line is broadened due to the diffusive atomic motion. The broad intensity maximum at shorter flight-times is due to energy gain processes on phonons (or on phonon-like excitations in the liquid). This is a direct indication for the existence of oscillatory motions in the liquid with frequencies similar to those in the solid.

Fig. 13 shows separately the quasielastic part of such spectra, after correction with regard to background and multiple scattering in the sample; furthermore, a transformation of the time-of-flight distribution into an energy transfer spectrum has been achieved. The dashed curves have been obtained from a fit with $S_{inc}(\kappa, \omega)$ according to the Egelstaff-Schofield model [52]; here $\gamma(t)$ has been taken from Eqs. (80) and (84) after folding S_{inc} with the resolution curve [9]. The delay time $1/\hat{\eta}$ and the mean square amplitude of the atomic vibrations $\gamma_\infty = \langle u^2 \rangle$ have been

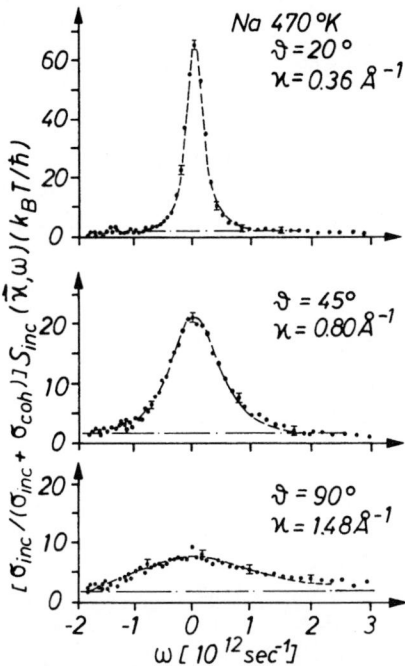

Fig. 13. Typical quasielastic lines for liquid sodium, after evaluation of the time-of-flight spectra (as in Fig. 12), in an absolute scale. Chain curve: "Inelastic background". Dashed line: Calculated from the Egelstaff-Schofield model (Section 4.2) with a best fit of the parameters $1/\hat{\eta}$ and γ_∞. κ = scattering vector; $\hbar\omega$ = energy transfer; σ_{inc}, σ_{coh} bound cross section (From Cocking [37])

used as disposable parameters. D is taken from macroscopic experiments. *Table 2* collects a few representative results.

Liquid argon has been studied by several authors [43, 75–77]. Zandveld et al. [75] have interpreted their experimental results in terms of the effective mass model discussed before. *Table 2* represents typical results for $\hat{M}/M$. Furthermore, the velocity autocorrelation function has been calculated by means of the quasi-crystalline model of Rahman

---

[9] Na scatters coherently and incoherently. At small κ, however, the incoherence dominates because the structure factor is small.

Table 2. Parameters of the diffusive motion in various liquids

Element	Temperature [°K]	$D \cdot 10^5$ [cm² sec⁻¹]	$\hat{\tau}_r \cdot 10^{12}$ [sec]	$\hat{M}/M$	$\langle r_\infty^2 \rangle = 6\gamma_\infty$ [Å²]	Ref.
Na	388	4.78	1.1	31	1.2	[37]
	470	8.40	1.5	29	1.9	
(solid)	367				1.0	
Pb	670	2.8	3.5	29		[37]
Ar	86	1.6	0.25	~ 3		[75]
	116	5.0	0.20	~ 1		

et al. [54] discussed in Section 4.2 *(Fig. 14)*; for this calculation the model parameters have been evaluated by fitting the theoretical scattering law with the experimental data. The points in Fig. 14 are taken from computer experiments.

After all that can be seen from real and from computer experiments, the velocity auto-correlation function has the following general features. After a steep initial decay within several 10^{-13} sec, there appears a negative overshoot region, due to the restoring force produced by the "cage" of the surrounding neighbours. As time goes on, this cage undergoes some random deformation; consequently, the correlation with the

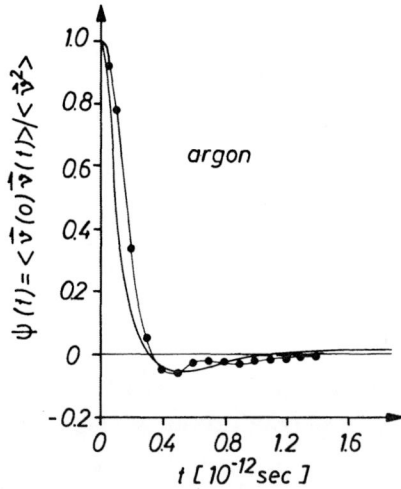

Fig. 14. Solid line: Normalized velocity auto-correlation function $\psi(t)$ as a function of time t for liquid argon (86° K) according to the theory of Rahman *et al.* [54] (Section 4.2), with model parameters taken from a best fit of scattering experiments by Zandfeld *et al.* [75]. Full circles: Computer experiments (94° K), from Rahman [14]. The Langevin theory would give $\psi = \exp\{-(k_B T/MD)\,t\}$.

initial velocity is being lost with a characteristic decay time of several 10^{-12} sec. For liquid metals [71], probably due to their longrange part of the potential, it seems that correlations with the neighbours exist for longer times than in noble gases; consequently, the oscillatory behaviour is more pronounced. This difference can be inferred directly from the observation that the spectra of liquid argon do not show pronounced inelastic peaks, in contrast to liquid sodium. Furthermore, this difference is also confirmed by the significantly different delay times and effective masses as shown in Table 2.

The *asymptotic behaviour* of $\langle v(0)\, v(t) \rangle$ in the region where the effects of the restoring force have decayed completely ($t > 5 \cdot 10^{-12}$ sec) is not yet well understood. From computer experiments it appears [74] that, for very large times, this function exhibits a positive tail; this means that the direction of the initial motion persists for an unexpectedly long time. From investigations of the space distribution of $v(t)$, sort of a vortex-flow pattern has been observed in the vicinity of the moving atom; this could be responsible for the above effect. Hydrodynamic calculations have led to conclusions in the same direction [74, 78]; they predicted an asymptotic decay law $\psi \sim t^{-3/2}$ which gives a $t^{1/2}$-term in $\langle r^2(t) \rangle$. An experimental confirmation of this behaviour [79] needs very accurate experiments at relatively small κ and ω.

5. Jump Diffusion in Liquids

In Section 4 the motion of the atoms in liquid noble gases and metals has been described by a simultaneous superposition of diffusive and oscillatory components. In certain liquids, the atoms are strongly associated with their next neighbours, for instance by hydrogen bonds as in water (see Section 5.2) or by short-living covalent bonds as in liquid tellurium [80]. In such cases, another description might be adequate. It resembles, to some extent, the diffusion model for atoms in a solid, as depicted in Fig. 10c: It is assumed that the atoms are consecutively trapped by such bonds in pseudo-equilibrium positions, within a certain time interval τ_0, during which they perform oscillations; then they jump or diffuse until they are trapped again in a new equilibrium position [81].

In quantitative terms, a description by such a model would apply if the height of the (fluctuating) potential barrier around the particle is sufficiently large compared to the characteristic oscillation frequency of the diffusing particle, i.e. $V_{\max} \gg \hbar\Omega_0$. Otherwise the diffusion is rather of the "continuous" type.

5.1 Theoretical Models

We describe the formalism for a jump diffusion model which directly leads to the selfcorrelation function $G_s(r, t)$, following the derivation given by Singwi and Sjölander [82]. The theory will be presented in some detail since the formalism is applicable for many other problems to be described later.

The motion of the particle is subdivided into successive steps $0, 1, 2, \ldots n$ (Fig. 10); we assume that the particle starts at the origin $r = 0$ for $t = 0$ being in its oscillatory state for a certain time interval which is τ_0 in the average (step 0). During the next step, 1, the particle is diffusing with a diffusion constant D_1 during an average time τ_1. In step 2 it is oscillating again, and so on. Starting from $r = 0$, $t = 0$, the particle could have arrived at r in time t after it has made $j = 0, 1, 2, \ldots$ steps. In this sense, the self-correlation function can be written as

$$G_s(r, t) = \sum_{j=0}^{\infty} F_j(r, t). \tag{100}$$

The $F_j(r, t)$ can be deduced in the following way. We call $g(r, t)$ the probability of finding the particle at r after a time t, if it has been performing an oscillatory motion around an equilibrium position, having started at $r = 0$ for $t = 0$. The probability that it remains in its oscillatory state within time t, is

$$p(t) = \exp\{-t/\tau_0\}. \tag{101}$$

For $g(r, t)$ we use the Gaussian form (30)

$$g(r, t) = [4\pi\gamma(t)]^{-3/2} \exp\{-r^2/4\gamma(t)\} \tag{102}$$

with an oscillatory width $\gamma(t)$ as in (81). Then one gets

$$F_0(r, t) = g(r, t)\, p(t). \tag{103}$$

The probability that the particle has left the oscillatory state in a time inverval between t and $t + \mathrm{d}t$, to start its diffusive motion, is

$$p(t) - p(t + \mathrm{d}t) = -(\mathrm{d}p/\mathrm{d}t)\,\mathrm{d}t = -\dot{p}(t)\,\mathrm{d}t = \exp\{-t/\tau_0\}\,(\mathrm{d}t/\tau_0). \tag{104}$$

For the diffusive motion we write, as in (78)

$$h(r, t) = [4\pi D_1 t]^{-3/2} \exp\{-r^2/4D_1 t\}. \tag{105}$$

The probability that the particle remains in the diffusive state is

$$q(t) = \exp\{-t/\tau_1\}. \tag{106}$$

Therefore one obtains

$$F_1(r, t) = \int_0^t (\mathrm{d}t_1/\tau_0) \int h(r - r_1, t - t_1) \exp\{-(t - t_1)/\tau_1\} \tag{107}$$

$$\cdot g(r_1, t_1) \exp\{-t_1/\tau_0\}\, \mathrm{d}r_1$$

Proceeding along these lines one finds the term $2n$ as a $2n$-fold convolution integral in space and time, namely

$$F_{2n}(r, t) = \int_0^t dt_{2n} \int_0^{t_{2n}} dt_{2n-1} \ldots \int_0^{t_2} dt_1 \int \ldots \int d\mathbf{r}_{2n} d\mathbf{r}_{2n-1} \ldots d\mathbf{r}_1 p(t - t_{2n}) \tag{108}$$

$$\cdot g(\mathbf{r} - \mathbf{r}_{2n}, t - t_{2n})\, \dot{q}(t_{2n} - t_{2n-1})\, h(\mathbf{r}_{2n} - \mathbf{r}_{2n-1}, t_{2n} - t_{2n-1}) \ldots \dot{p}(t_1) g(\mathbf{r}_1, t_1).$$

The scattering law is the space-time Fourier transform of (100) with (108). After introducing new time variables $t - t_{2n} = \tau_{2n+1}, t_{2n} - t_{2n-1} = \tau_{2n}$ etc., and analogous conventions for the space coordinates, the integrals in the sum (100) can be factorized yielding a simple geometrical series. Leaving out numerous intermediate steps of the calculation, one finds

$$S_{\text{inc}}(\boldsymbol{\kappa}, \omega) = \frac{\tau_0}{\tau_0 + \tau_1} \frac{A[1 + (B/\tau_0)]}{1 - (AB/\tau_0\tau_1)} + \frac{\tau_1}{\tau_0 + \tau_1} \frac{B[1 + (A/\tau_1)]}{1 - (AB/\tau_0\tau_1)}. \tag{109}$$

Here A and B are the space-time Fourier transforms of the functions $p(t) g(\mathbf{r}, t)$ and $q(t) h(\mathbf{r}, t)$, respectively. The appearance of *two* terms in (109) takes into account that only a certain fraction of the particles has started with an oscillatory motion (as has been tacitly assumed above); the other particles have started in their diffusive state. Therefore, $G_s(r, t)$ had to be averaged over both, with weight factors

$$\tau_0/(\tau_0 + \tau_1) \quad \text{and} \quad \tau_1/(\tau_0 + \tau_1),$$

respectively. It is important that (109) holds for any motion consisting of two kinds of successive steps with mean duration τ_0 and τ_1, if A and B are defined in a proper sense.

Introducing $\langle l^2 \rangle = 6\tau_1 D_1$, the mean square displacement during time τ_1, one finds the macroscopic selfdiffusion constant from the asymptotic behaviour of $G_s(r, t)$; one obtains the relation

$$D = \tfrac{1}{6}(\langle l^2 \rangle + 6\langle u^2 \rangle)/(\tau_0 + \tau_1). \tag{110}$$

During the derivation of (109) it has been assumed that no correlation exists between the motions in consecutive steps. This is justified if the particle performs, in the average, a great number of oscillations until it diffuses to another quasi-rest position; this means

$$\tau_0 \gg 1/\Omega_0, \tag{111}$$

where Ω_0 is the average frequency of the particle. Obviously, vibrations with a frequency smaller than $1/\tau_0$ cannot develop. For $\tau_0 \lesssim 1/\Omega_0$, the separation into oscillatory and diffusive steps looses its sense.

In the following, we restrict ourselves to the discussion of a few limiting cases of (109). For a very strongly associated liquid the particles are in the bound and oscillatory state for most of their time which means $\tau_0 \gg \tau_1$. In this case, $\langle l^2 \rangle$ is just the mean square jumping distance. The scattering law is then found to be a Lorentzian, namely

$$S_{\text{inc}}(\kappa, \omega) = \frac{(\Gamma/2\hbar\pi)\exp\{-\kappa^2\langle u^2\rangle\}}{\omega^2 + (\Gamma/2\hbar)^2}.$$ (112)

The width at half-maximum is then given by

$$\Gamma = 2\hbar \frac{\kappa^2 D + (1 - \exp\{-\kappa^2\langle u^2\rangle\})/\tau_0}{1 + \kappa^2 D\tau_0}.$$ (113)

For $\kappa^2 D\tau_0 \simeq \kappa^2\langle l^2\rangle/6 \ll 1$ this leads to [10]

$$\Gamma = 2\hbar\kappa^2[D + (\langle u^2\rangle/\tau_0)] \simeq 2\hbar\kappa^2 D.$$ (114)

Assume now that $\kappa^2\langle l^2\rangle \gg 1$. In this case one obtains

$$\Gamma = 2\hbar/\tau_0.$$ (115)

    This asymptotic form has been derived very early by Brockhouse [48] under the assumption that the particle performs jumps between quasi-lattices sites where it spends an average time τ_0. One should notice that the width (115) should be interpreted as a lifetime effect of the oscillating atom only in the following sense: "Lifetime" means the *residence time of the oscillating atom within the volume of size $1/\kappa^3$ "seen" during the scattering process.*

    If one assumes a broad statistical distribution of the jump lengths l as $l\exp\{-l/\langle l\rangle\}$, one can show that the jump model yields [44]

$$\Gamma = 2\hbar\kappa^2 D/(1 + \kappa^2 D\tau_0).$$ (116)

This gives the same asymptotic behaviour as described by (114) and (115). Finally, Eq. (109) reproduces purely "continuous" diffusion (Eq. (78)) for $\tau_1 \ll \tau_0$. No Debye-Waller factor appears in this case.

    So far, it has been assumed that the particle, if "trapped" in its oscillatory state, has a center of oscillation which is fixed in space; actually it undergoes a slow diffusive motion. A theory has been worked out [83] which takes this into account; it is a synthesis between a step model with pronounced crystalline features as discussed in this section, and the model of continuous diffusion from Section 4. As a result, the quasi-elastic width does not saturate for large κ-values as it occurs in (113) and (116).

---

[10] Except for $u^2 \ll l^2$ there appears a certain inconsistency as (113) does not yield the proper behaviour for $\kappa \to 0$ which should be $\Gamma = 2\hbar\kappa^2 D$. However, the criterium $u^2 \ll l^2$ should be fulfilled in any case; otherwise the supposition of well-defined rest positions and of independent diffusive steps would not hold at all.

5.2 Experiments on Water

No liquid has been investigated by so many authors as water, including a great number of studies by means of neutron spectroscopy [48–51, 53, 84–89, 94]. Numerous models concerning the structure of this liquid have been proposed which were mainly based on evidence from X-ray diffraction and from optical measurements. Most of them agree on the point that part of the molecules are connected with their neighbours

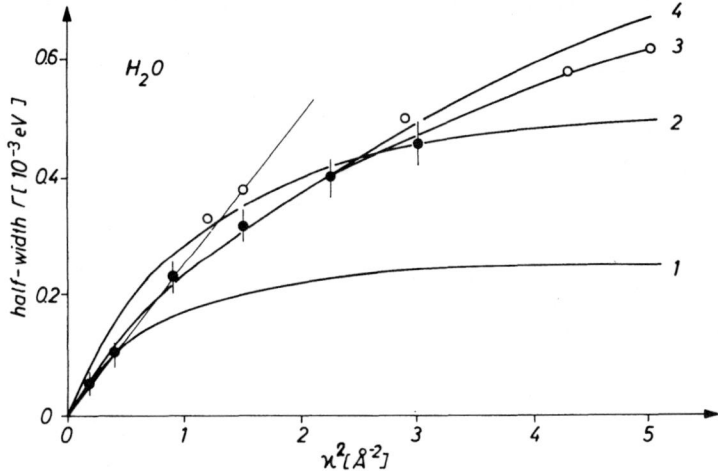

Fig. 15. Full width at half-maximum of quasielastic line vs. κ^2 for water at room temperature. Full circles: Franks *et al.* [88]; open circles: Safford *et al.* [87]; curves (1) and (2): Jump models (113) with $\tau_0 = 5 \cdot 10^{-12}$ sec and $\tau_0 = 2.5 \cdot 10^{-12}$ sec, respectively. (3) Jump model with continuous distribution of the step lengths (116) with $\tau_0 = 1.2 \cdot 10^{-12}$ sec. (4) Model of continuous diffusion, according to Eqs. (80)–(84) with $\hat{\tau}_r = 4.7 \cdot 10^{-12}$ sec. Straight line calculated with $D = 1.9 \cdot 10^{-5}$ cm^2/sec (Eq. (79))

by strong hydrogen bonds; the assumed arrangement is similar to that in the structure of ice (see for instance [90]). It is to be expected that these ice-like bonds have a certain lifetime during which adjacent molecules vibrate against each other. For the rest of their time the molecules are diffusing or jumping freely. In this picture, a step model as discussed before might allow an adequate interpretation of neutron scattering experiments.

Optical and neutron scattering spectra show the existence of several pronounced *inelastic peaks*. They have been ascribed to translational and torsional vibrations of the molecule against its quasi-crystalline neighbourhood (for instance [51]). In addition, the neutron spectrum includes the *quasielastic line* which can easily be separated from the inelastic spectrum up to κ-values of ~ 2 Å^{-1}. *Fig. 15* presents typical results for the quasielastic width Γ plotted as a function of κ^2, together with theoretical calculations.

For small κ, all experiments show the linear behaviour of Γ vs. κ^2 as predicted by (114), and the agreement of D with the values determined by macroscopic methods is better than 10% [91]. Such an agreement should exist for a monoatomic liquid but it is not necessary for molecules due to the influence of rotations [88] (see Section 8). From the agreement stated before one might conclude that, for water, the rotational motions are *so fast* that they produce a flat background which does not affect the experimental line width at all; otherwise they might be *so slow* that the molecule performs many translational steps before it rotates. Nuclear magnetic resonance experiments [92, 93] suggest that the time during which molecular orientation persists $(3...8 \cdot 10^{-12}$ sec) is several times larger than the rest time τ_0 to be discussed subsequently.

For large κ the width curve bends downwards. In this region there is some controversy concerning the experimental data. For part of them, Γ seems to approach a constant value [51] which would agree with the prediction of the Singwi-Sjölander model (113); in other experiments Γ vs. κ^2 increases steadily. As can be inferred from Figure 15 it is possible to interpret the data either with the modified jump model (116) (assuming a distribution of jump lengths) or with the model of continuous diffusion as discussed in Section 4. In Section 5.3 some aspects will be discussed by which one could discriminate between such alternatives.

The pronounced quasi-crystallinity of H_2O, as it manifests itself by strong peaks in the inelastic spectrum suggests an interpretation of the quasielastic intensity in terms of a Debye-Waller factor as in (112). From *Fig. 16* a mean square amplitude of the oscillating proton $\langle u^2 \rangle = 0.34 \text{ Å}^2$ is being obtained. Also this number shows some discrepancy with results of other authors who have found smaller values. For a comparison between the measured $\langle u^2 \rangle$ and the calculations on the basis of the vibrational spectrum of the H_2O molecules we refer to the literature [89].

An analysis of the existing data in the frame of the jump models (113) or (116) gives, as disposable parameter of the model, the average rest time; this has a value between 1 and $2 \cdot 10^{-12}$ sec. The periods of molecular vibrations are concentrated in the region of 10^{-13} sec. Therefore, a molecule performs, in the average, at least ten vibrations until it leaves its pseudolattice site. This would justify the application of the model. On the other hand, from this model the average jump distance can be calculated to be about 1 Å. This means that the criterium $l^2 \gg u^2$ is fulfilled rather poorly (see [88]). As a whole, results and interpretation seem to be somewhat ambiguous, in spite of the great number of investigations existing so far. This is partly due to the fact that a decision between the various models needs data in the region of larger κ-values. These are difficult to obtain because there the quasielastic line is weak as compared to the "background" of inelastic and other processes. This statement holds in general, and careful measurements at large κ-values

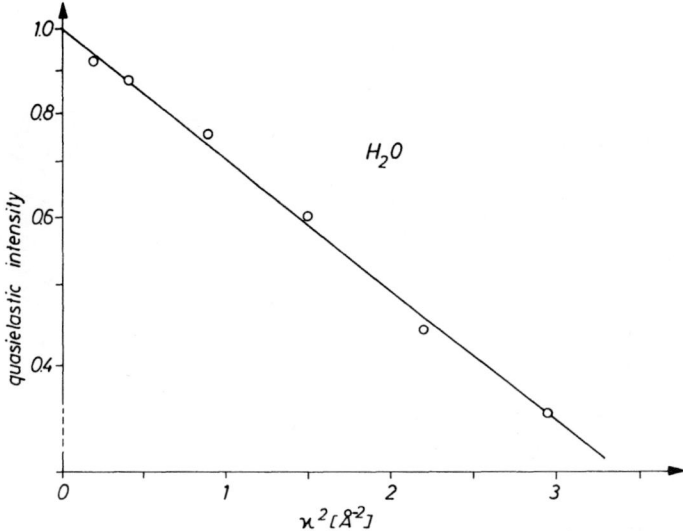

Fig. 16. Intensity of the quasielastic line of water (log-scale). Solid line: Effective Debye-Waller factor of the vibrating H_2O molecules, $\exp\{-\kappa^2\langle u^2\rangle\}$, with $\langle u^2\rangle = 0.34\,\text{Å}^2$ (From Franks et al. [88])

with higher energy resolution are necessary to allow an analysis of the shape, and not only of the width of quasielastic lines.

Investigations on *heavy water*, where complications arise in the interpretation due to coherence effects, will not be reported (see for instance [95, 96]).

5.3 Difference between Continuous and Jump Diffusion

From the results of different models which are summarized in a qualitative manner in *Fig. 17*, several general conclusions can be drawn.

(i) For small κ-values, a Lorentzian-shaped quasielastic line is obtained in all cases. For a monoatomic liquid its width is $\Gamma = 2\hbar\kappa^2 D$, where D is the macroscopic self-diffusion constant.

(ii) For intermediate κ-values $(1\ldots3\cdot\text{Å}^{-1})$ the curve Γ vs. κ^2 bends strongly, and the width and the shape of the quasielastic line are determined by the atomistic features of the diffusive process.

(iii) For liquids with quasi-crystalline behaviour, i.e. for $\tau_0 \gg 1/\nu_D$, τ_1 or $\omega'/\omega_D \ll 1$, more or less pronounced peaks occur in the inelastic spectrum. They correspond to the typical vibration frequencies of the diffusing particles. In such cases, a separation of the quasielastic line from the inelastic spectrum is feasible, and the mean square amplitude of the particle vibration $\langle u^2\rangle$ can be determined from the area of the quasielastic line.

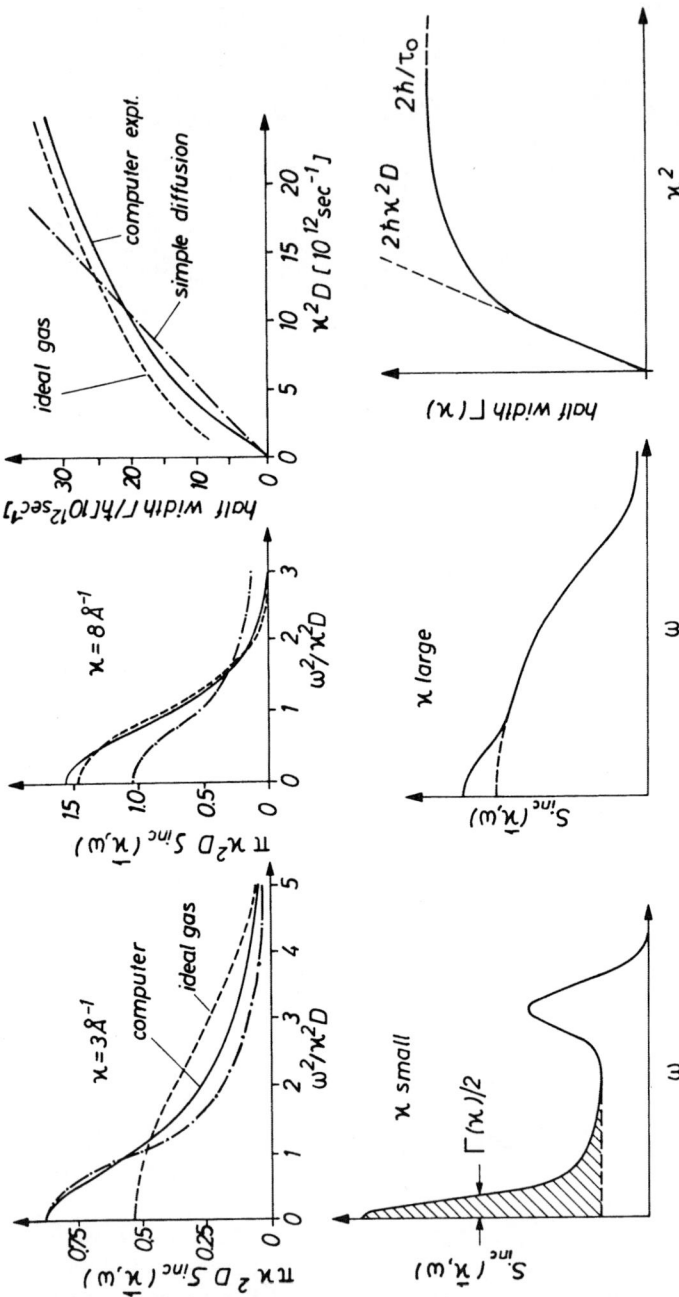

Fig. 17. Survey on the general features of the incoherent scattering law $S_{inc}(\kappa, \omega)$ and of the full width at half-maximum of the spectrum, $\Gamma(\kappa)$. First row: Liquid argon. Solid line: Width function $\gamma(t)$ taken from computer experiments with $G_s(\mathbf{r}, t)$ in Gaussian approximation. Dashed line: Ideal gas, see (117). Dot-dashed: Simple diffusion with $\gamma(t) = Dt$, so that $S_{inc}(\kappa, \omega)$ is Lorentzian-shaped as in (78). From Nijboer and Rahman [17]. Second row: Liquid with pronounced crystalline features like water (schematic sketch). The quasielastic line is separable from the "inelastic background" (dashed) if κ is small. $\Gamma(\kappa)$ is the full width at half-maximum of the *separated* quasielastic spectrum. The shaded area is proportional to the Debye-Waller factor of the liquid. τ_0 = average rest time. The peak at large ω corresponds to vibrational states of the diffusing atom. For a solid, the quasielastic line had to be replaced by a δ-function

On the other hand, for continuous diffusion without a pronounced oscillatory component, the scattering law does not contain a Debye-Waller factor; this is due to the fact that all degrees of freedom are taking part in the diffusive motion.

Additional information can be obtained from the known self-diffusion constant. Combining it with the various models one obtains ω'/ω_D, $M/\hat{M}$, or the average jump length.

(iv) For large $\kappa(>3...4\,\text{Å}^{-1})$ the quasielastic line practically disappears (if it was separable at all), and merges with the inelastic spectrum. The scattering law then approaches a Gaussian as it would be obtained for an ideal gas *(free gas limit)*, namely [58]

$$S(\kappa, \omega) = (2\pi\Delta^2)^{-1/2} \exp\{-(\omega - \hbar\kappa^2/2M)^2/\Delta^2\} \tag{117}$$

with $\Delta^2 = 2\kappa^2\langle E_{\text{kin}}\rangle/3M$ which approaches $\kappa^2 k_B T/M$ for high temperatures. In this region, the scattering law is insensitive with respect to the details of the atomic motions.

In view of what has been said it is easy to understand the information to be obtained from experiments at small and at large κ-values. For *small* κ, the contributions to the Fourier integral (22) originate from a large space volume of size $1/\kappa$. Therefore, the scattering process "observes" the motion over long paths, which means over many diffusive steps. This leads to (114) which does not depend on the nature of the single step. On the other hand, for *large* κ (in the jump model $\kappa > 1/l$) the neutron wave packet interacts only along a single step of the diffusion process. Therefore, the scattering law depends, for instance, on τ_0 or τ_r.

It should be emphasized that the neutron *scattering* process and the line width cannot be understood classically by a recoil effect for a collision with an atom having a certain momentary velocity. This should be distinguished from the fact that the *motion* of the scattering particle can, under certain circumstances, very well be calculated in terms of classical physics.

Finally, we deal with the possibility of a discrimination between continuous and step diffusion. We compare the corresponding formulas for D, namely

$$D = \tau_r k_B T/\hat{M} \quad \text{"continuous"} \tag{118}$$

$$D = \tfrac{1}{6}\langle l^2\rangle/\tau_0 \quad \text{"jump"}. \tag{119}$$

Apparently, the characteristic times enter in a reverse way. This is easily understood since D *increases* if the average time τ_r increases in which the velocity vector keeps its memory; on the other hand, D *decreases* if the mean time of stay τ_0 increases.

By observing the width Γ plotted versus $\kappa^2 D$ as a function of T for fixed $\kappa^2 D$ one can discriminate between both types of models because of the opposite temperature dependence according to Eqs. (85) and (115). Such considerations [53] were found to be in favour of the continuous model for liquid metals. For water no clear decision has been possible. This indicates that the applicability of a jump model is an oversimplification, even for a liquid of such strong intermolecular interactions as water.

6. Diffusion of Hydrogen in Metals

In a number of metals, hydrogen can be dissolved with very high concentrations where the hydrogen atoms occupy sites of an interstitial lattice (see *Fig. 18*). In certain regions of the phase diagrams, these hydrogens are extremely mobile, and jump rates as high as 10^{12} per second

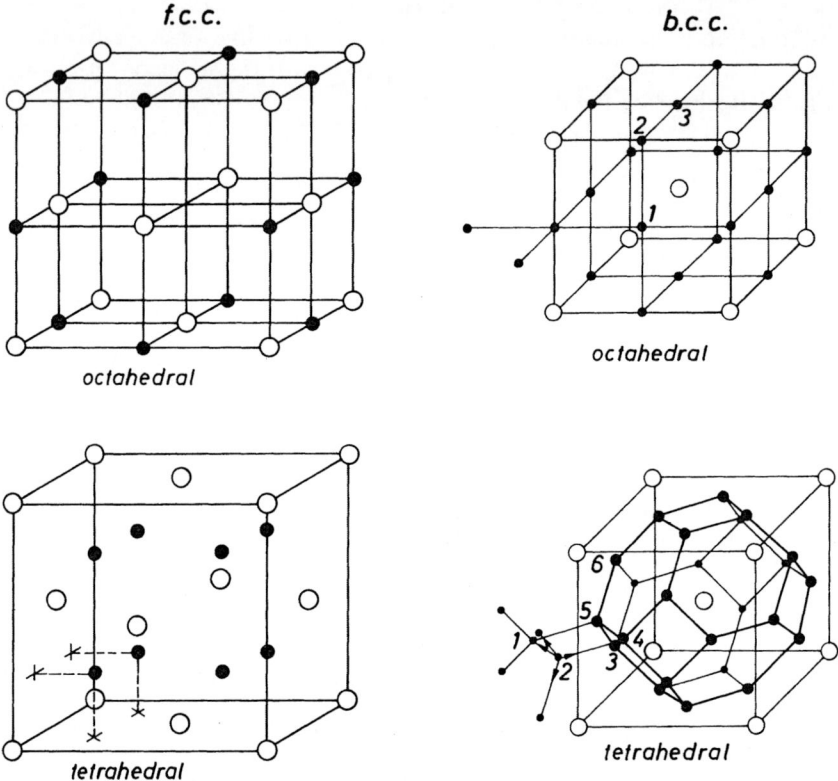

Fig. 18. Various types of interstitial sites in fcc and bcc host lattices (full circles). Open circles: host lattice atoms. The non-equivalent interstitial sites for the bcc lattice are labelled with 1,2,3 (octahedral) and with 1 ... 6 (tetrahedral)

have been observed. This fact, in connection with other observations, leeds to the conclusion that, under certain conditions, hydrogen in metals behaves like a *"lattice gas"*, or a *"lattice liquid"*, where the hydrogen is trapped in the cage formed by the host lattice atoms [97].

Quasielastic neutron scattering on these protons is a method to obtain information concerning these fast protonic motions; for instance, one gets the jump rate, the self-diffusion coefficient, the jump geometry, and the shape of the thermal cloud of the proton vibrating around its interstitial site. Because the incoherent cross-section of the proton is very high compared to that of the metals it is relatively easy to correct with regard to scattering on the host lattice.

6.1 Theoretical Models

For the calculation of the self-correlation function we use, as in Section 5, a jump model, supposing that the diffusion process consists of thermally activated steps which are statistically independent. This assumption is likely to be justified if the proton performs many oscillations between successive jumps. Furthermore, we neglect the time τ_1 needed for a jump from site to site, compared to the mean time of stay τ. For a single step this means

$$\tau \gg \tau_1 \simeq s(k_B T/2M)^{-1/2} \simeq 10^{-13} \text{ sec} \tag{120}$$

where s is the jump distance. Finally, for the sake of simplicity, we assume that the presence of the protons at neighbouring sites does not affect the jump probability. This may hold even for relatively high atomic concentrations as there are many interstitial positions available per metal atom (see Fig. 18).

In contrast to jump diffusion in a liquid, we have to deal with quasi-equilibrium sites which are forming a *periodic interstitial lattice*. The probability $P(r, t)$ of finding a proton at a point r of this lattice at a time t should then follow a rate equation (Chudley and Elliott [98]), namely

$$\partial P(r, t)/\partial t = (1/n\tau) \sum_{k=1}^{n} \left[P(r + s_k, t) - P(r, t) \right]. \tag{121}$$

Here, τ is the mean time of stay at the rest position; $P(r + s_k)$ is the probability of finding a proton in a neighbouring site; $s_k (k = 1, ..., n)$ are the jump vectors leading to the neighbouring sites. Per definition one has then

$$G_s^D(r, t) \equiv P(r, t), \quad \text{if} \quad P(r, 0) = \delta(r) \tag{122}$$

is imposed as initial condition. $G_s^D(r, t)$ is the diffusive part of the self-correlation function. In (121) it has been tacitly assumed that the set of jump vectors, s_k, is identical for all interstitial sites; this means that they form a simple Bravais lattice. As can be seen from figure 18, this holds for fcc, but not for bcc host lattices (see below).

Now we introduce the intermediate scattering law in the following general form (compare with Eq. 37)

$$I_s^D(\kappa, t) = \exp\{-t f(\kappa)\} \tag{123}$$

so that

$$G_s^D(r,t) = (2\pi)^{-3} \int I_s^D(\kappa, t) \exp\{-i\kappa r\} \, d\kappa. \tag{124}$$

With (121) to (124) one obtains

$$f(\kappa) = (1/n\tau) \sum_{k=1}^{n} (1 - \exp\{-i\kappa s_k\}). \tag{125}$$

For a simple cubic lattice this gives

$$f(\kappa) = (1/3\tau)(3 - \cos\kappa_x\alpha - \cos\kappa_y\alpha - \cos\kappa_z\alpha) \tag{126}$$

where α is the cubic lattice parameter. The Fourier transformation (19) of (123) leads to a simple Lorentzian

$$S_{inc}^D(\kappa, \omega) = \frac{f(\kappa)/\pi}{\omega^2 + f^2(\kappa)} \tag{127}$$

with a full-width at half-maximum

$$\Gamma = 2\hbar f(\kappa). \tag{128}$$

For $\kappa \to 0$, (126) approaches $\Gamma = 2\hbar\kappa^2(\alpha^2/6\tau) = 2\hbar\kappa^2 D$.

For interstitial lattices which are not of the Bravais type, the calculation of the scattering law is complicated. For instance, in the case of a *bcc* host lattice there exist in one unit cell $m = 3$ non-equivalent sites with an octahedral, and $m = 6$ sites with a tetrahedral arrangement of the neighbouring host lattice atoms (Fig. 18).

As a consequence, we have to introduce sets of jump vectors $s_{ij,k}$ which connect a site of type $i(i = 1, ..., m)$ with a neighbouring site of type $j(j = 1, ..., m)$; k labels these jump vectors. Calling $P_i(r_l, t)$ the probability to find the hydrogen, for time t, at a site of type i in a cell l (with position vector r_l), and assuming that the jump probability $1/\tau$ for all inequivalent sites is equal, one gets a system of m rate equations [99, 100]

$$\partial P_i(r_l, t)/\partial t = (1/n\tau) \sum_{j,k} [P_j(r_l + s_{ij,k}, t) - P_i(r_l, t)]. \tag{129}$$

The sum has to be taken over all n neighbouring sites j of the site i in cell l. Here the initial condition reads

$$P_i^{(v)}(r_l, 0) = \delta_{iv}\delta_{l0} \tag{130}$$

where δ are Kronecker symbols. This means that the hydrogen should be for $t = 0$ at a site of type v within the cell at the origin, $l = 0$. Consequently, the P_i according to (129) are forming a set of self-correlation functions $G_{vi}^D(r_l, t) \equiv P_i^{(v)}(r_l, t)$; they describe the probability of finding the proton for time t at a site of type i within cell l, if it has started at a site v within cell 0 at $t = 0$. The actual $G_s^D(r, t)$, as needed for the calculation of $S_{inc}^D(\kappa, \omega)$, is then an average over all v and a sum over all i. As a consequence of the existence of more than one interstitial site per cell, the scattering law is a superposition of a number of Lorentzians rather than a single one as in (127).

An alternative mathematical approach has been worked out by Gissler *et al.* [101]; it is advantageous for complicated interstitial lattices, and for the treatment of systems with multiple jumps as they might occur at elevated temperatures. As in Section 5, we write

$$G_s^D(r, t) = \sum_{j=0}^{\infty} F_j(r, t) . \tag{131}$$

The terms F_j can be factorized as

$$F_j(r, t) = w_j(r)\, p_j(t) \tag{132}$$

where w_j is the probability that a particle, starting at $r = 0$, could have arrived at r after it has performed j jumps. $p_j(t)$ is the probability that j jumps have occured during the time t, namely

$$p_j(t) = (t/\tau)^j\, (j!)^{-1}\, \exp\{-t/\tau\} . \tag{133}$$

Obviously one has $w_1(r) = \sum_{i=1}^{n} \delta(r - s_i)$ and consequently

$$w_j(r) = n^{-j} \sum_{p} \delta[r - (s_1 + s_2 + \cdots + s_j)] . \tag{134}$$

The sum has to be taken over all possible paths "p", each composed of a combination of j jumps starting at $r = 0$[11]. Applying the formalism of random walk theory [46], recursion formulas have been developed to calculate the w_j step by step with a computer until the convergence of the sum (131) is satisfactory.

Fig. 19 shows the half-width $\Gamma(\kappa)$ of the quasielastic line as it has been determined by such calculations for cases where jumps are taking place between adjacent octahedral or between adjacent tetrahedral sites in a *bcc* host lattice.

As a consequence of the periodic arrangement of the interstitial sites, the width $2\hbar f(\kappa)$ is a periodic function of κ. For all κ-values approaching a reciprocal vector $\mathcal{G}$ of the interstitial lattice, the width becomes zero. It is easily recognized that the Fourier components $\exp\{-t f(\kappa)\}$ in (124) which do not decay for $t \to \infty$, are just originating from these κ-values [102]. We now expand κ around the reciprocal interstitial lattice vector $\mathcal{G}$.

---

[11] For the sake of simplicity the possibility of different starting sites has been disregarded here.

Writing $\kappa = \mathscr{G} + q$ we have $f(\mathscr{G} + q) = f(q)$; with (123) and (124) this leads to

$$\lim_{t \to \infty} G_s^D(r, t) = (2\pi)^{-3} \sum_{\text{all } \mathscr{G}} \int_{q \to 0} \exp\{-t f(\mathscr{G} + q)\} \exp\{-i(\mathscr{G} + q) r\} \, dq$$

$$\equiv (2\pi)^{-3} \sum_{\text{all } \mathscr{G}} \exp\{-i\mathscr{G}r\} \int_{q \to 0} \exp\{-t f(q) - i q r\} \, dq \,. \tag{135}$$

Here, $t f(q)$ approaches Dt for small t. Therefore, the right hand side of (135) reproduces the correct asymptotic form of $G_s^D(r, t)$, namely the interstitial lattice sum with a Gaussian-shaped envelope according to Eq. (30) with $\gamma = Dt$. In this connection it should be pointed out that, in general, a *discrete hopping distance* is able to produce a more or less pronounced

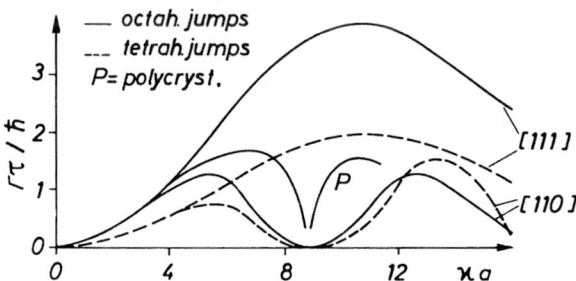

Fig. 19. Full width at half maximum Γ in units of the mean jump rate $1/\tau$ vs. scattering vector κ for jumps between interstitial sites of the octahedral and the tetrahedral type in a bcc lattice (Fig. 18), for various orientations of κ and for a polycrystal. $a = $ cubic host lattice parameter. $\Gamma\tau/\hbar = 2$ would be the asymptotic value for an isotropic jump model. Calculated by Gissler and Rother [101] with the random walk method

oscillatory behaviour of the quasielastic width (see Section 10.2). This effect is inherently connected with strong deviations from a Gaussian-shaped $G_s(r, t)$.

In addition to the diffusive motions, the protons perform vibrations around their interstitial sites. These consist of two components; one, where the proton follows the vibrations of the host lattice, and another which as a frequency ω_0 beyond the upper limit of the host lattice vibrations *(localized mode)*. Formally, these vibrations can be treated as in (80) and (93) by writing

$$\gamma(t) = t f(\kappa) + \gamma_{\text{osc}}(t) \tag{136}$$

if correlations are neglected.

The oscillatory part in $\gamma(t)$ is again responsible for a Debye-Waller factor which determines the quasielastic intensity[12]. The total mean square amplitude of the proton $\langle u^2 \rangle$ can be estimated in the harmonic approximation as a sum of the contributions from the localized and the host lattice vibrations (see Eq. (82)). This leads to

$$\langle u^2 \rangle = (\hbar/2 M_P \omega_0) + 3 g k_B T / M_h \omega_D^2 \tag{137}$$

assuming $\hbar \omega_0 \gg k_B T$. Here M_P and M_h are the masses of the proton and host lattice atom, respectively.

---

[12] As a consequence of the fact that the vibrations contain localized modes, the Debye-Waller factor deviates from its usual form [103–105]; quantitatively, however, this modification is negligible in most cases.

The contribution due to the host lattice has to be weighted by an unknown factor $g = \langle u_P^2 \rangle / \langle u_h^2 \rangle$. It takes into account that the proton amplitudes u_P are not necessarily the same as those of the host lattice, u_h. For hydrogen in vanadium, g has been determined experimentally being of the order of 1,5 [104]. Extensive calculations of the frequencies and the mean-square amplitudes for impurities can be found in the literature (for instance [106]; for *bcc* lattices see [107]).

6.2 Experiments

So far, the diffusive motion of hydrogen has been studied in the systems Pd—H [108, 109], Nb—H [110, 111], and V—H [100]. For the α-phase of fcc Pd—H, Sköld and Nelin [108] have been the first who observed

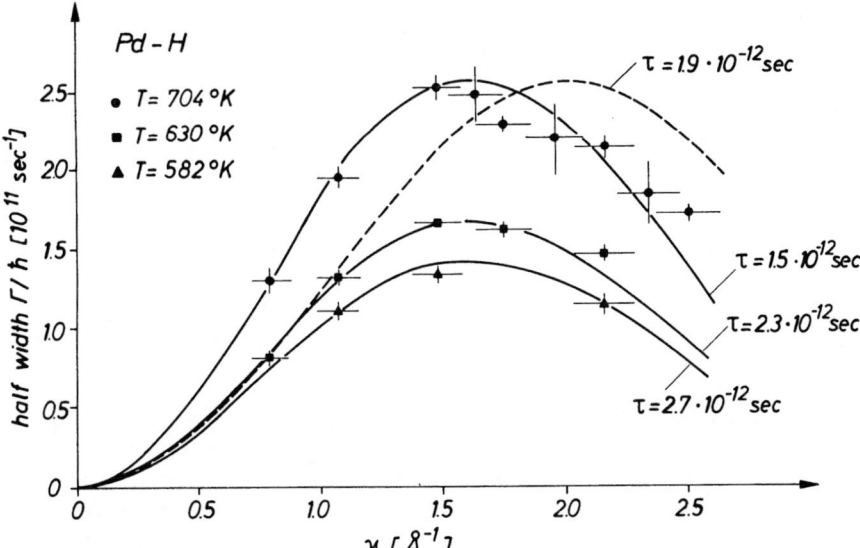

Fig. 20. Half-width of the quasielastic line for scattering on hydrogen dissolved in poly-crystalline palladium at various temperatures (concentrations between 2 and 4 atomic percent). Theoretical calculations using the theory of Chudley and Elliott [98] for jumps between octahedral sites (solid line) and tetrahedral sites (dashed line); see Fig. 18. τ = mean rest time at the sites, according to a best fit of the curves (From Sköld and Nelin [108])

that the width of the quasielastic line has a maximum. *Fig. 20* shows their experimental results. The curves in the figure have been calculated according to the theory of Chudley and Elliott (Section 6.1) for models with jumps between octahedral, and between tetrahedral sites (see Fig. 18). Because only polycrystalline samples were available, the scattering law (and not $f(\kappa)$ in Eq. (123)) had to be averaged over all directions in space. It can be seen that the curves $\Gamma(\kappa)$ for the octahedral

Table 3. Selected results from quasielastic neutron scattering on hydrogen in metals

Sample	temperature	self-diffusion constant D (cm²/sec)[a]	rest time τ (sec)	activation energy of D or $1/\tau$ E_{act}	mean-square amplitude of proton $\langle u^2 \rangle$[d]	Ref.
α-PdH$_{0.02-0.04}$	582° K	$4.6 \cdot 10^{-5}$ $(3.3 \cdot 10^{-5})$	$2.7 \cdot 10^{-12}$	0.16 eV	–	[108]
	704° K	$8.4 \cdot 10^{-5}$ $(7.8 \cdot 10^{-5})$	$1.5 \cdot 10^{-12}$ (o–o jumps)		0.13 Å²	
β-PdH	373° K	$1.06 \cdot 10^{-5}$	$13 \cdot 10^{-12}$ (o–o and o–t jumps)	0.14 eV		[109]
Nb–H$_{0.33}$	[b] 393° K $(< T_c)$	$0.8 \cdot 10^{-5}$	–	0.12 eV		[111]
	583° K $(> T_c)$	$2.7 \cdot 10^{-5}$	$2 \cdot 10^{-12}$ (for o–o jumps)		0.18 Å²	
α-Nb–H$_{0.03}$	393° K	$2.0 \cdot 10^{-5}$ $(1.9 \cdot 10^{-5})$	–			
α-V–H$_{0.2}$ [c]	483° K	$8 \cdot 10^{-5}$ $(10 \cdot 10^{-5})$	$0.7 \cdot 10^{-12}$ (for t–t jumps	0.047 eV	0.18 Å²	[100]

[a] Results from macroscopic measurements in brackets.
[b] $(\alpha + \alpha')$-phase.
[c] No observable broadening in β-phase.
[d] Directional average because the experiments have been performed on polycrystals.

and for the tetrahedral case have their maximum at different values of κ. A clear decision in favour of a diffusion model with jumps between octahedral sites has been possible.

Experiments on the β-phase of Pd—H [109] could not be interpreted in terms of such a simple model. Assuming a mixture of octahedral-octahedral and octahedral-tetrahedral jumps, the proportion of both could be estimated. The analysis turns out to be even more ambiguous for *bcc* crystals. Here the periodicity of $\Gamma(\kappa)$ is the same for the octahedral and the tetrahedral case (Fig. 19), in spite of the fact that the jump distances differ considerably[13].

Table 3 compares a selection of quantitative data obtained from existing quasielastic scattering experiments, namely D from the slope of Γ vs. κ^2 for small κ and the rest time τ from a fit with the theory for a certain jump model; furthermore Table 3 shows the activation energy E_{act} as obtained from fitting the T-dependence of D with an Arrhenius equation $D = D_0 \exp\{- E_{act}/k_B T\}$. The values of D and E_{act} for Nb—H from neutron scattering agree quite well with those determined by means of the relaxation process caused by long range diffusion of the protons

---

[13] The periodicity of $f(\kappa)$ depends on the translation vectors of the reciprocal interstitial lattice, and not on the jump vectors.

[112]. The quantity $1/\tau$ has also been compared with the results of nuclear magnetic resonance experiments [113]; they are, however, quite unreliable in this region because $1/\tau$ is several orders of magnitude higher than the *nmr* frequencies.

It should be realized that in all these systems, jump rates and self-diffusion constants for the protons have been found which are of the same order of magnitude as those observed for atoms in ordinary liquids! On the other hand, in spite of the smallness of τ, the proton can perform quite a number of oscillations between successive jumps; therefore, pronounced peaks due to localized modes are observable. It should be emphasized that, if τ becomes very short, the jumping time τ_1 is no longer negligible, and it is to be expected that the corresponding scattering law approaches that of a free gas.

A point of special interest is the behaviour of the protonic motion in the vicinity of a *phase transition*. For V—$H_{0.57}$, a drastic change of the quasielastic width occurs in passing from the high temperature bcc-α-phase, through the intermediate $(\alpha + \beta)$ phase, to the tetragonally distorted β-phase; here the quasielastic width becomes unobservably small [100]. This might be interpreted on the basis of a change in the lattice structure: The tetragonal β-phase leaves more space for a certain kind of the interstitial sites than it is available in the undistorted α-phase; this might lead to a stronger binding of the protons sitting on these sites of the β-phase. In fact, nuclear magnetic resonance experiments on the β-phase indicate a higher activation energy for proton jumps than it has been found from neutron scattering in the α-phase.

In the systems Nb—H or Pd—H, the phase transitions are of quite different nature. The striking similarity of the phase diagrams with those of a real gas suggests, as already mentioned, the existence of a "protonic lattice gas" in the α-phase, and of a "lattice liquid" in the α'-phase; the boundary of the miscibility gap then corresponds to the coexistence curve, having a critical point on its top. Below T_c, a mixed phase $(\alpha + \alpha')$ exists which corresponds to the coexistence of lattice gas and lattice liquid. The behaviour of the quasielastic line has been studied across the critical point of Nb—H [111]. The ratio of the diffusion constant (from Γ for small κ) over the value at some large κ has been extracted from the experiments; this ratio has the meaning of a square jump length. It has been found that this ratio undergoes a strong change near T_c. For $T > T_c$ one could conclude that jumps occur between octahedral sites, while for $T < T_c (\alpha + \alpha'$-phase) tetrahedral jumps should be dominating. The latter statement must, however, be taken with caution because of the heterogenity of this phase.

A subject of considerable interest is the discussion of the *mean square amplitude* $\langle u^2 \rangle$ for the vibrating proton (Table 3). It turns out that, in general, this quantity is several times larger than the values estimated according to (137) if one uses an average amplitude ratio of $g = 1$. In particular, experiments on Nb—H have demonstrated an anomalous increase of $\langle u^2 \rangle$ in the region of the critical temperature *(Fig. 21)*. Between 130° and 170° C it amounts to about 40% of the value at 300° C. No change has been observed in the inelastic spectrum which would be compatible with this result.

Obviously, the motion of hydrogen in metals has a number of exceptional features, essentially due to the small mass and size of the proton; we will further comment on this point, in particular with respect to the *bcc* metals: Protons at octahedral sites might have a very large extension of their vibrational wave function in directions perpendicular to the axes which connect the nearest-neighbour host atoms. This might lead to an especially high chance of motions in the spacious [100] directions of the interstitial lattice. In this connection it is noteworthy

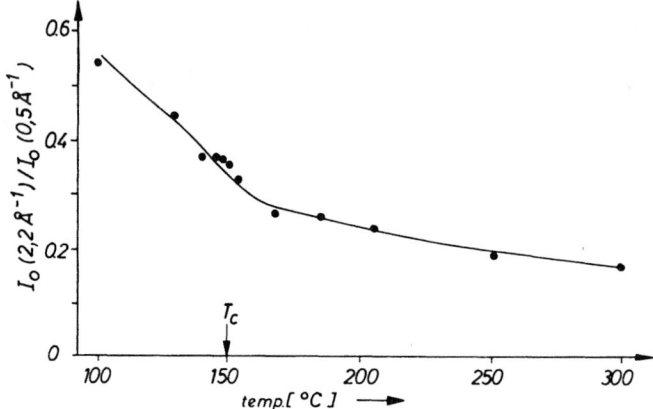

Fig. 21. Intensity ratio of the quasielastic line in $NbH_{0.33}$ for small and large values of κ. This ratio is approximately proportional to the Debye-Waller factor. An anomalous decrease is observed for $T \approx T_c$ = critical temperature of the protonic "lattice gas" (From Gissler et al. [111])

that the energies of the localized frequencies in Nb—H [110], found at $\hbar\omega_0 \simeq 0.11$ eV and 0.18 eV, are equal or even higher than the energy of activation for self-diffusion, $E_{act} = 0.11$ eV [112]. This is an indication that the first excited level of the proton lies in the continuum, suggesting the idea of a bandtype motion of the protons [110]. At low temperatures, D has been found to decrease more slowly with decreasing temperature as predicted by an Arrhenius law [112]. This might be attributed to a failure of the picture which describes diffusion as a classical rate process [114–116]. The diffusion constant in that region is of the order of 10^{-6} cm^2/sec; consequently, quasielastic scattering, which might give information on the kind of the diffusion process, would still be feasible.

7. Rotational Diffusion in Molecular Solids

In most cases, molecules in crystals have an orientational degree of freedom and they perform reorientational motions due to thermal excitation. These motions can occur between indistinguishable and/or between distinguishable orientations (see *Figs. 22 and 23*). If two or more distinguishable orientations exist, the crystal exhibits disorder with respect to the molecular orientations. In such cases, phase transitions can occur where the degree of disorder undergoes a change [118].

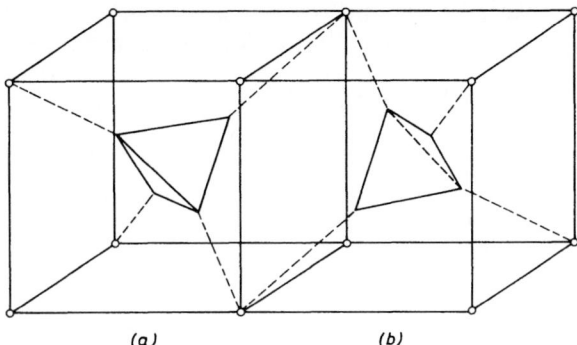

(a)   *(b)*

Fig. 22. Two distinguishable orientations of a molecule with tetrahedral shape in a crystal like NH_4Cl. The orientations can be transformed one into the other by 90° rotations around a twofold axis. Jumps between *indistinguishable* orientations of the molecule occur around the threefold axes with a 120°-rotation

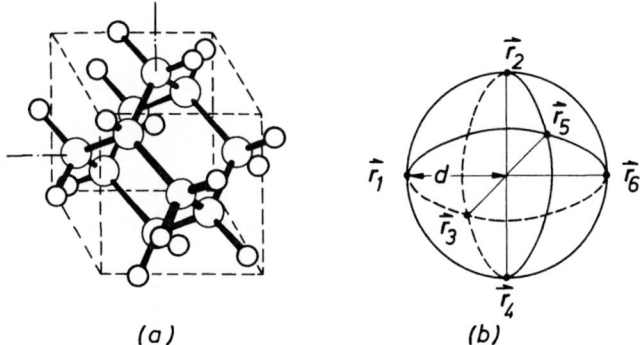

(a)   *(b)*

Fig. 23. a Adamantane molecule ($C_{10}H_{16}$). Considering for simplicity only the motion of the proton pairs sticking out of the cube, and replacing them by a single one, one obtains a 90°-jump geometry as shown in b

A particular situation exists in the so-called plastic phases of certain organic crystals [119] consisting of molecules with globular shape. Here the orientational disorder is large. Consequently, the rotational motion might be a random-walk diffusion of the molecular axes without well-defined quasiequilibrium directions, resembling the molecular motion in a liquid.

A description in terms of well-defined quasiequilibrium positions will probably be adequate for systems like the ammonium halides (at least in their low temperature phases), where the rotating NH_4-groups are well separated by the rigidly bound halogen-ions. The "liquid-like" description of rotation is supposed to be more adequate in the high-temperature phases of organic crystals where the rotating molecules are in direct "touch" with each other; as a consequence, a strong coupling

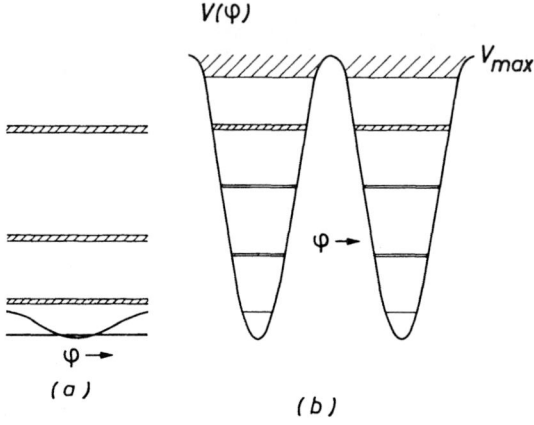

Fig. 24. Nearly free (a) and strongly hindered rotation (b) of a molecule in a crystal. $V(\varphi)$ is the angle dependent potential of hindrance (schematic); φ is the rotation angle. The energy levels of the rotating molecule have a finite width due to interactions with thermal lattice waves

exists between them. So far, no conclusive decision between a discrete jump and a continuous diffusive motion has been possible in such cases, and quasielastic scattering could help to clarify these questions. Nearly *free* rotation is found very seldom, for instance in solid hydrogen where it leads to collective rotational waves [120], and probably in the high-temperature phase of NH_4Cl around one of the axes of the NH_4-tetrahedra [121].

To formulate the distinction in more quantitative terms [118], we speak of "*nearly free*" rotation in cases where the energy states of the rotating molecule $E_0, E_1, \ldots$ are far above the maximum of the orientational potential, V_{max}; therefore *(Fig. 24)*

$$\Delta E > V_{max} . \tag{138}$$

ΔE is the level distance. Under these circumstances, also the disturbancy of the rotation due to thermal fluctuations of the hindrance potential is weak; thus the rotation is able to persist for quite long time intervals,

τ', so that the levels have a small width. This means that

$$\hbar/\tau' \ll \Delta E. \tag{139}$$

The opposite case is a *strong hindrance* potential, so that

$$E_0, \Delta E \lesssim V_{\text{max}}. \tag{140}$$

If, in addition, the librational states are relatively sharp, the concept of "*jump rotation*" might be applicable, i.e. if for the "rest time" τ

$$\hbar/\tau \ll \Delta E. \tag{141}$$

The case of reorientational jumps will be treated theoretically in Section 7.1, assuming discrete quasiequilibrium orientations determined by the lattice geometry.

On the other hand, "*rotational diffusion*" would mean that the rotational levels are completely blurred by interactions between the rotations and the thermal lattice vibrations. This type of motion will be treated in Section 7.2 by means of the well-known *orientational correlation functions*.

Finally we discuss how to visualize the motion which the molecule performs *between* two quasi-equilibrium orientations. At lower temperatures, this motion is supposed to be a "jump"; this means that the molecular axis takes the shortest path from one equilibrium orientation to the other, and the time needed for this motion is given by a relation similar to (120). Such jump models might be applicable for jump rates lower than $10^{12} \sec^{-1}$. There might also be cases where the molecular axis, after having left its equilibrium orientation, rotates or diffuses freely for quite a long time until it is trapped in a new equilibrium orientation. This will be treated at the end of Section 7.2, assuming quasi-equilibrium orientations which are distributed isotropically.

Under certain circumstances, the rotations might have the character of *tunneling transitions*, and the concept of a classical jumping time looses its meaning. This case might be observable experimentally if the splitting of the librational levels (due to the periodical potential) is larger than the width of these levels due to thermal vibrations of the lattice.

The *scattering law* for the rotating molecules shows a shape as sketched in *Fig. 4a* (Section 2). If the molecule performs a random rotational motion, there exists a bell-shaped quasielastic line; its width is determined by the time constant of this rotational motion. If the motion is a free rotation, the spectrum should reveal side-maxima; these might merge due to damping mechanisms[14]. Because the proton is bound

---

[14] According to our definition in Section 1, such *periodical* motions do not produce quasielastic scattering. Here we will, however, include them because a rigorous separation of free rotation and rotational diffusion is artificial.

to a molecule which practically cannot leave its lattice position, the self-correlation function $G_s(r, t)$ remains finite for $t \to \infty$. As a consequence, the spectrum shows, in additon to the quasielastic part, a sharp elastic line at $\hbar\omega = 0$, as discussed generally in Section 2.3. Its area gives information on the *average space distribution* of the rotating proton. In principle, this line, too, would have a small width due to translational diffusion of the whole molecule; however, in solid crystals this width is too small to be observable in neutron spectra. The typical rotational times which could be observed experimentally are, according to Fig. 6, in a region from 10^{-8} to 10^{-12} sec.

It should be pointed out that in most theories correlations between rotational and translational motions will be completely neglected. This might be a good approximation for more or less spherical molecules, but rather poor for molecules of elongated shape. This coupling would be an interesting problem in itself (see [19]).

7.1 Jump Models with Equilibrium Orientations Determined by Crystal Symmetry

To calculate $G_s(r, t)$ we use the following simplified model (Stockmeyer *et al.* [122], Sköld [123]). The scattering proton, being rigidly bound to a molecule in a crystal, alternates between librational motions around certain quasi-equilibrium positions r_ν ($\nu = 1, ..., z$), and rapid rotations between these positions. We assume that the average time τ, for which the proton librates, is large compared to the time τ_1 needed to rotate from one equilibrium orientation to the other.

First we calculate the self-correlation function $G_s^{(\mu)}(r, t)$ for a proton, which was for $t = 0$ at a distinct equilibrium position, namely r_μ; we have shifted r_μ into the origin, so that $r_\nu = 0$ for $\nu = \mu$. The actual correlation function later on is obtained by performing an average over all possible choices of r_μ[15] (because of $\tau_1 \ll \tau$ we neglect starting positions *between* the various r_ν). This leads to

$$G_s^{(\mu)}(r, t) = \sum_{\nu=1}^{z} g(r - r_\nu, t)\, p_\nu^{(\mu)}(t). \tag{142}$$

Here $g(r - r_\nu, t)$ is the time-dependent self-correlation function for a librational motion of the proton in the molecule around one of the quasi-equilibrium positions r_ν. It could be formulated in analogy to Eqs. (102) and (81).

$p_\nu^{(\mu)}(t)$ is the probability of finding the proton in the librational state at r_ν, if it was, at an earlier time $t = 0$, at the selected origin $r_\mu = 0$ (the set

---

[15] This procedure corresponds, for $t \to \infty$, to the formalism given in general terms in Section 2.3.

of vectors r_ν differs for different μ). Then one gets for the actual self-correlation function

$$G_s(r, t) = (1/z) \sum_{\mu=1}^{z} G_s^{(\mu)}(r, t) \tag{143}$$

assuming that all positions are equivalent. If successive jumps are uncorrelated, the $p_\nu^{(\mu)}$ can be calculated from a set of rate equations in analogy to (129), namely

$$d p_\nu^{(\mu)}(t)/dt = (1/n\tau) \sum_{\lambda=1}^{n} [p_\lambda^{(\mu)}(t) - p_\nu^{(\mu)}(t)] \ (\nu = 1, \dots z). \tag{144}$$

Here the sum goes over all those n positions r_λ from which the proton can jump directly to position r_ν. The resulting $p_\nu^{(\mu)}(t)$ are exponential decay functions which obviously fulfill the relations

$$p_\nu^{(\mu)}(0) = 1 \quad \text{for} \quad \nu = \mu \quad \text{and} \quad \sum_{\nu=1}^{z} p_\nu^{(\mu)} = 1 \quad \text{for all } \mu. \tag{145}$$

The "stationary" or equilibrium distribution is then

$$p_\nu^{(\mu)}(\infty) = 1/z \quad \text{for all } \nu \text{ and } \mu. \tag{146}$$

Performing the Fourier transform (22b) of (142) one finally obtains the incoherent scattering law.

We explicitly demonstrate the calculation of $S_{\text{inc}}(\kappa, \omega)$ for the case of a hypothetical dumb-bell molecule with scattering protons on both ends. It is supposed to perform rotational jumps around an axis perpendicular to the molecular axis *(Fig. 25)*. We need

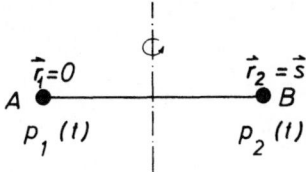

Fig. 25. For the derivation of Eq. (150): Dumb-bell molecule with two scattering protons, performing rotational jumps around the dashed axis. P_1 and P_2 are the probabilities to find a proton at r_1 and r_2, respectively, if it has started at r_1 for $t = 0$

only to consider one of the protons, say A. The occupancy of site (1) by A, if it has started at (1), is

$$d p_1^{(1)}(t)/dt = -(1/\tau) p_1^{(1)}(t) + (1/\tau) p_2^{(1)}(t) \tag{147}$$

with the initial condition $p_1^{(1)}(0) = 1$. This leads to

$$p_1^{(1)}(t) = \tfrac{1}{2}(1 + \exp\{-2t/\tau\}) \equiv p(t), \tag{148}$$

by using

$$p_2^{(1)}(t) = 1 - p(t). \tag{149}$$

For simplicity we replace $g(r - r_v, t)$ by $\delta(r - r_v)$. This corresponds to a suppression of the librational motion; we then obtain, with $r_1 = 0$, $r_2 = s$, the expression

$$G_s^{(1)}(r, t) = \delta(r) \, p_1^{(1)}(t) + \delta(r - s) \, [1 - p_1^{(1)}(t)] \, . \tag{150}$$

The probability $p_1^{(2)}$ for the case that proton A has started at site (2), is then found by interchanging the indices. This leads to $G_s(r, t)$, the average of both cases, namely

$$G_s(r, t) = \delta(r) \, p(t) + \tfrac{1}{2} \delta(r - s) \, [1 - p(t)]$$
$$+ \tfrac{1}{2} \delta(r + s) \, [1 - p(t)] \, . \tag{151}$$

Fourier transformation in space and time finally gives

$$S_{\text{inc}}(\kappa, \omega) = \tfrac{1}{2} \delta(\omega) \, (1 + \cos \kappa s) + \frac{(1/\tau \pi) \, (1 - \cos \kappa s)}{\omega^2 + (2/\tau)^2} \, . \tag{152}$$

The intensity of the purely elastic part decreases as κ increases, whereas the quasielastic part behaves in the opposite way, the sum being unity as it should be.

In certain cases it has to be noticed that the average time between successive jumps of a single proton in a molecule, τ, is not necessarily identical with the average time between reorientational jumps of the whole molecule, τ_M. For instance, rotational jumps of a proton bound to a CH_4-molecule [123] through 120° around the threefold axes (Fig. 22) evidently occur between indistinguishable orientations of the molecule; these leave always one of the four protons unaffected. Therefore one gets $\tau = (4/3) \, \tau_M$. An extensive discussion of such systems of rate equations for rotational jumps has been worked out in connection with the interpretation of nuclear magnetic resonance experiments (for instance [121]).

Introducing also the *oscillatory motions* of the proton in the molecule leads to an inelastic contribution in the spectrum; furthermore, the spectrum has to be multiplied by a Debye-Waller factor. The latter can be strongly anisotropic due to the fact that the mean square amplitude of the center of gravity, $\langle u_c^2 \rangle$, in general differs from the librational square amplitude $\langle u_l^2 \rangle$ [117]; the latter is inversely proportional to the moment of inertia of the molecule. Therefore, the "proton cloud" has approximately the shape of a rotational ellipsoid with a thickness u_c and a diameter u_l. The resulting anisotropy of the Debye-Waller factor is superimposed with the anisotropy due to the stationary behaviour of $G_s(r, t)$ discussed before. For small κ, both functions start as $(1 - \text{const } \kappa^2)$; only at larger κ-values the typical features of the structure factor of the jumping proton become visible. Unfortunately, measurements at large κ are difficult because the "background" of inelastic processes is large due to phonons and librons.

7.2 Description by Orientational Correlation Functions

An alternative way of describing the rotational motion is to apply rotational correlation functions. This is, in principle, more general than the rate process formalism in Section 7.1. For this purpose we calculate the incoherent scattering law via the intermediate scattering function

$I_s(\kappa, t)$ from the time-dependent space coordinates of the moving proton, $r_i(t)$ (see Eq. 20b). We introduce the position vector $R(t)$ of the center of gravity of the scattering molecule, and the distance vector of the scattering proton within this molecule from its center of gravity, $d_i(t) = r_i(t) - R(t)$. Scattering on other atoms than protons will not be considered. We follow the derivation given by Sears [124] where rotational-translational correlations are neglected; this allows a simple factorization of (20b) into a translational and a rotational part, namely

$$I_s(\kappa, t) = I_s^{(t)}(\kappa, t) I_s^{(r)}(\kappa, t)$$
$$= \sum_i \langle \exp\{i\kappa[R(t) - R(0)]\}\rangle \langle \exp\{i\kappa[d_i(t) - d_i(0)]\}\rangle . \tag{153}$$

The sum goes over all protons within one molecule. We now define a *rotational correlation function* as an analogue of the van Hove correlation function [124]:

$G(\Omega, \Omega_0, t - t') \, d\Omega$ is the probability of finding *the orientation of a molecule in a solid angle element* $d\Omega$ *centered at the orientation* Ω *at time t, given that the orientation was* Ω_0 *at an earlier time* t'. Ω and Ω_0 may be characterized by sets of Eulerian angles.

This correlation function is of general importance for the description of all kinds of interaction processes on molecules [125, 126]. With its help, the average $\langle\ldots\rangle$ in the rotational part of (153) can be formulated as follows

$$I_s^{(r)}(\kappa, t) = \langle \exp\{i\kappa[d(t) - d(0)]\}\rangle = (1/8\pi^2) \int G(\Omega, \Omega_0, t)$$
$$\cdot \exp\{i\kappa[d(t) - d(0)]\} \, d\Omega \, d\Omega_0 . \tag{154}$$

This expression considers scattering only from a single proton. It holds for all types of motion as long as a classical description is allowed.

To simplify the formalism it will be assumed that the rotation is *isotropic*. This means that rotations through the same angle about different space axes occur with the same probability. This holds for spherical top molecules if the potential field acting on the molecule is isotropic.[16] Under these circumstances, the function $G(\Omega, \Omega_0, t)$ and the exponential in the integrand of (154) can be expanded into spherical harmonics. Using the condition that the orientations are centered at Ω_0 for $t = 0$ and that $\int G(\Omega, \Omega_0, t) \, d\Omega = 1$, one obtains, with the help of the orthogonality relations and the addition theorem for spherical harmonics

$$I_s^{(r)}(\kappa, t) = \sum_{l=0}^{\infty} (2l + 1) j_l^2(\kappa d) F_l(t) . \tag{155}$$

---

[16] For a polycrystal it should be emphasized that the directional average has to be taken on $S_{inc}(\kappa, \omega)$ and not on $G_s(r, t)$.

The $j_l(\kappa d)$ are spherical Bessel functions. The expansion coefficients are the *orientational autocorrelation functions*, namely

$$F_l(t) = \int P_l[\cos\beta(t)]\, G(\boldsymbol{\Omega}, \boldsymbol{\Omega}_0, t)\, d\boldsymbol{\Omega} \equiv \langle P_l[\cos\beta(t)]\rangle \tag{156}$$

Here $\beta(t)$ is the angle through which a vector, fixed to a certain point of the molecule, rotates within a time interval t[17]. P_l is a Legendre polynomial. The coefficients with $l = 1$ and 2 are being used in the theory of optical spectroscopy and of nuclear magnetic resonance (Section 11).

If the molecular axis performs a random motion, the orientation looses its memory with increasing time. As a consequence, all $F_l(t)$ decay to zero, except $F_0 = 1$.

Fourier transformation of (155) gives the scattering law

$$S_{\mathrm{inc}}(\boldsymbol{\kappa}, \omega) = j_0^2(\kappa d)\,\delta(\omega) + \sum_{l=1}^{\infty} (2l+1)\, j_l^2(\kappa d)\,\tilde{F}_l(\omega) \tag{158}$$

where the Fourier transformed correlation functions are

$$\tilde{F}_l(\omega) = (1/2\pi) \int \exp\{-i\omega t\}\, F_l(t)\, dt\,. \tag{159}$$

The expansion coefficients $\tilde{F}_l(\omega)$ can be calculated on the basis of various models which describe the motion of the molecular axis in terms of classical physics [127]. For an entirely *free rotation* of a spherical molecule in a crystal one obtains for $l > 0$ [124]

$$\tilde{F}_l(\omega) = (2l+1)^{-1}\,[\delta(\omega)$$
$$+ (4/\sqrt{\pi}) \sum_{\xi=1}^{l} (I/2\xi^2 k_B T)^{3/2}\,\omega^2 \exp\{-I\omega^2/2\xi^2 k_B T\} \tag{160}$$

where I is the moment of inertia of the molecule.

Assume, on the other hand, that $G(\boldsymbol{\Omega}, \boldsymbol{\Omega}_0, t)$ satisfies a *diffusion* equation for the orientational motions [193, 124]; this means

$$D_r \Delta_{\tilde{\Omega}}\, G(\boldsymbol{\Omega}, \boldsymbol{\Omega}_0, t) = \partial G(\boldsymbol{\Omega}, \boldsymbol{\Omega}_0, t)/\partial t \tag{161}$$

where $\Delta_{\tilde{\Omega}}$ is the Laplacian for the rotations and D_r is the rotational diffusion constant. With the initial condition that the orientation $\boldsymbol{\Omega}$ equals $\boldsymbol{\Omega}_0$ for $t = 0$, one obtains

$$F_l(t) = \exp\{-l(l+1)\, D_r t\} \tag{162}$$

---

[17] The F_l can also be formulated in terms of the orientation angle of the molecule with respect to an axis fixed in space, namely (for instance [126])

$$F_l(t) = (2l+1)\langle P_l[\cos\theta(0)]\, P_l[\cos\theta(t)]\rangle \tag{157}$$

where $\theta(t)$ is the angle of the molecular axis with respect to an arbitrary axis fixed in space at a certain time t.

To obtain the proper behaviour at small t, this formula has to be modified in analogy to Eq. (74) in Section 4 [127, 128].

According to (162) the functions $\tilde{F}(\omega)$ are Lorentzian-shaped with a full-width at half maximum *(Fig. 26)*

$$\Gamma_l = 2\hbar l(l+1)\, D_r\,. \tag{163}$$

Therefore, contrary to the jump models of Section 7.1, the scattering law is a sum of an *infinite* number of Lorentzians.

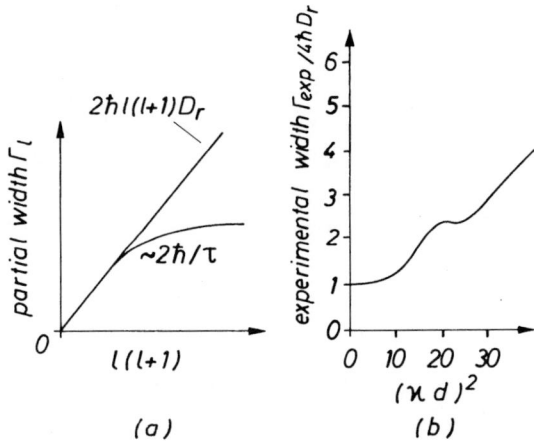

Fig. 26. a Full width at half-maximum of the expansion coefficients $\tilde{F}_l(\omega)$ of the quasi-elastic spectrum for diffusively rotating molecules, as a function of l (Eqs. (158) and (162)). b Experimental full width at half-maximum Γ_{exp}, i.e. the width of the composite spectrum (Eq. (158), including the $\delta(\omega)$-term), after broadening by a finite resolution width Δ where Δ is chosen to be $4\,\hbar D_r$, (d = distance of scattering nucleus from the center of gravity of the rotating molecule). Schematic, see [127]

Another case of physical importance is the *isotropic rotational jump model* [130]. Assume that the angles ε of rotational jumps are randomly distributed around a certain average value $\langle\varepsilon\rangle$, following a probability distribution proportional to $\exp\{-\varepsilon/\langle\varepsilon\rangle\}$ [127]; then one finds $\Gamma_l = 2\hbar l(l+1)\langle\varepsilon^2\rangle/6\tau$ for small l as in (163); for large l (under certain geometrical conditions not to be discussed here) the width approaches $\Gamma_l = 2\hbar/\tau$, where τ is the average time between the jumps *(Fig. 26)*.

Obviously, the first term of the expansion (155) leads always to a purely elastic line, $\delta(\omega)$, which follows from the finite asymptotic value of $G_s(r, t)$ (Section 2.3). Diffusive rotation leads, in addition to the elastic line, to a bell-shaped quasielastic spectrum centered at $\omega = 0$, whereas *free* rotation gives Maxwellian-shaped *side maxima*. In the latter case, also the terms $l > 0$ in (155) contain a non-decaying contribution, so that they give rise to delta-functions in ω. According to (160), the total intensity of the purely elastic part for the free rotations is then found to be

$$F_0(\kappa d) = \sum_{l=0}^{\infty} j_l^2(\kappa d) = (Si2\kappa d)/2\kappa d \tag{164}$$

whereas for the other cases one gets from (158)

$$F_0(\kappa d) = j_0^2(\kappa d) = (\sin \kappa d)^2/(\kappa d)^2 . \tag{165}$$

It should be noticed that the width of the quasi-elastic terms in (158) does not depend on κ, contrary to the behaviour of translational diffusion. On the other hand, the "*effective*" width at half maximum of a spectrum *measured with finite resolution* increases with κ. This follows from the fact that the intensity of the purely elastic contribution decreases in proportion to the quasielastic contribution. This holds for small κ. At higher κ-values, the effective width might show an oscillatory behaviour due to the spherical Bessel functions (see Fig. 26).

It is sometimes useful to rearrange the sum (155) as follows [128]

$$I_s^{(r)}(\kappa, t) = \exp\{-(\kappa^2 d^2/3)[1 - F_1(t)]\}[1 + O(\kappa d)^4 + \cdots] . \tag{166}$$

For small κ, the time-dependent higher terms $O(\kappa d)^4$ in the square bracket can be neglected, and (166) approaches the familiar Gaussian approximation (37), depending on the dipolar term $F_1(t)$ only.

Dahlborg et al. [131, 131a] have formulated the theory of rotational motions in more general terms, assuming that the molecular axis alternates between two kinds of motion, namely librations around quasi-equilibrium positions during an average time τ_0, and some kind of rotational motion during τ_1 as discussed above, both superimposed with the oscillations of the center of gravity of the molecule. Under these circumstances Eq. (109) can be applied, namely

$$S_{\text{inc}}(\kappa, \omega) = \frac{\tau_0}{\tau_0 + \tau_1} \frac{a[1 + (b/\tau_0)]}{1 - (ab/\tau_0\tau_1)} + \frac{\tau_1}{\tau_0 + \tau_1} \frac{b[1 + (a/\tau_1)]}{1 - (ab/\tau_0\tau_1)} , \tag{167}$$

with

$$a(\kappa, \omega) = \int_0^{\infty} \exp\{-i\omega t - (t/\tau_0)\} I_s^{(t)}(\kappa, \omega) I_s^{(\text{lib})}(\kappa, t) \, dt , \tag{168}$$

$$b(\kappa, \omega) = \int_0^{\infty} \exp\{-i\omega t - (t/\tau_1)\} I_s^{(t)}(\kappa, \omega) I_s^{(r)}(\kappa, t) \, dt . \tag{169}$$

Here $I_s^{(t)}$, $I_s^{(\text{lib})}$, and $I_s^{(r)}$ are the intermediate scattering functions for vibrations of the center of mass, for librations, and for rotational motions of the molecular axis, respectively.

The finite "life time" of the scattering particle in one or the other "state" of motion has lead to the introduction of decay factors $\exp\{-t/\tau_0\}$ or $\exp\{-t/\tau_1\}$ in Eqs. (167)–(169). After Fourier transformation, they are responsible for the appearance of Lorentzians in the scattering law. This means that this formalism yields a finite width of the spectral line at $\hbar\omega = 0$, even if the center of gravity of the rotating molecule is fixed in the crystal. This obviously contradicts the general arguments in Section 2.3 according to which a sharp elastic line should be expected in this case. Such a contradiction appears also in the theory mentioned at the end of Section 8.1. In the meantime, the origin of this difficulty has been clarified, and a modified formulation of the problem will be discussed in a forthcoming paper by Larsson [201]. The essential results remain unchanged; however, the quasielastic line appearing in (167) has to be replaced by a sharp line of the same intensity.

7.3 Experiments

The most serious difficulty in the investigation of orientational motions in crystals consists in the separation of the sharp line from the rotational quasielastic spectrum, and in the separation of the latter from the

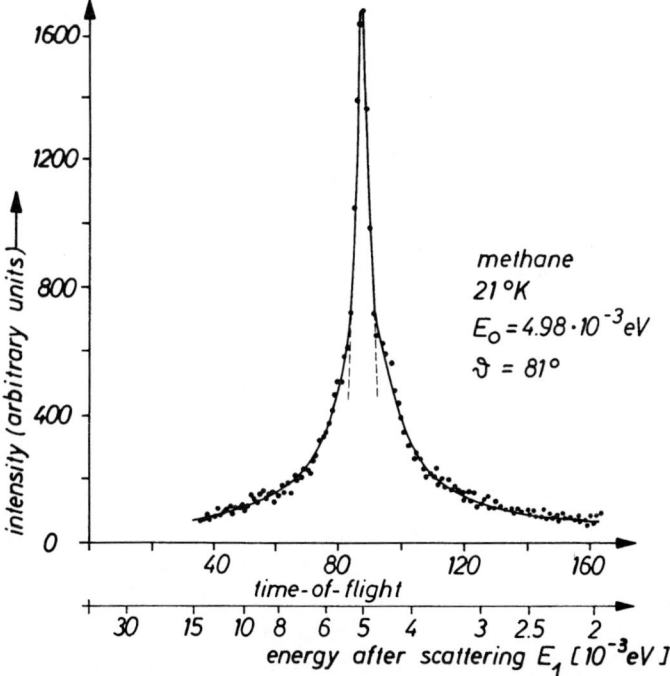

Fig. 27. Time-of-flight spectrum on solid methane in its high temperature phase. Scattering angle ϑ; incident energy E_0. A separation is possible between a quasielastic component and a component with a width which agrees with the instrumental resolution (From Kapulla [132])

"background" produced by low-energy phonon scattering. Only a few experiments have been able to succeed in such a separation.

Fig. 27 shows, as an example, a spectrum for *solid methane* in its high-temperature phase [132]; both spectral components can be seen separately due to a fairly high jump rate. From neutron diffraction studies on CD_4 single crystals [133] it is known that the methane tetrahedra occupy in the average 12 equivalent equilibrium orientations, with some finite probability of orientations distributed between them. Asserting that the "sharp" component of the spectrum is the (resolution-broadened) purely elastic line, and interpreting the quasielastic part simply in terms of jumps between the orientations, a rate of the order of $1/\tau \simeq 10^{12}$ sec^{-1} has been found.

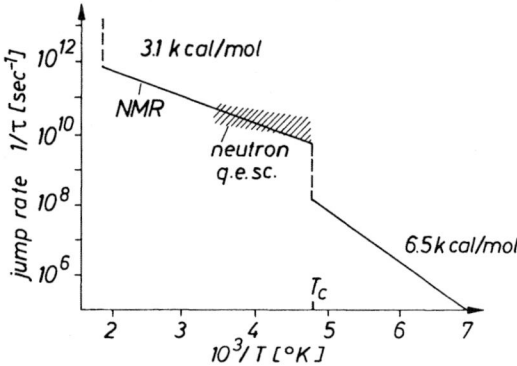

Fig. 28. Rotational jump rate of an adamantane molecule as a function of temperature T as determined from a least square fit of neutron spectra with a theory using a rotational jump model (shaded); from Stockmeyer and Stiller [122]. Solid line: Nuclear magnetic resonance data (Resing [134]). Transition into the plastic phase at T_c. Activation energies are indicated

In *adamantane* ($C_{10}H_{16}$), considerable scattering intensity has been observed in time-of-flight spectra below 10^{-3} eV, if the crystal is transformed into its plastic high-temperature phase [122]. The scattering law according to a jump theory in analogy to Eq. (152), convoluted with the known resolution curve of the spectrometer, has been fitted with the measured spectra; from this one gets the jump rate $1/\tau$ as a function of temperature. The calculations for $S_{inc}(\kappa, \omega)$ have been carried out on the basis of 90°-rotations around the cubic axis with discrete orientations, for the sake of simplicity neglecting scattering from the corner protons of the molecule (see Fig. 23). In *Fig. 28* the results are compared with jump rates determined by means of nuclear magnetic resonance, showing fairly good agreement. It might well be that the experimental results can also be interpreted in terms of other models for the molecular rotation. Experiments on single crystals with a larger κ-range will give a better insight into this question.

Further experiments have been reported by Lechner *et al.* on *plastic neopentane* [135]; again a fit has been performed with a spectrum calculated by means of a jump model with discrete orientations, in order to obtain the jump rate and its activation energy. On the same substance, experiments by other authors [136] have been interpreted in terms of a combined rotational diffusion and a rotational jump motion. However, the model they have used does not properly take into account that the proton is bound to a molecule fixed in the crystal. A clear separation of elastic and quasielastic scattering has not been possible in these experiments.

Scattering experiments across the *ferroelectric* phase transition of $(NH_4)_2SO_4$ have revealed, in addition to the line at $\hbar\omega = 0$, a quasielastic contribution [131]. The results have been interpreted in terms of an alternating rotational and librational motion of the NH_4-ions as discussed at the end of section 7.2. From a measurement of the intensity of the sharp elastic peak as a function of κ it has been possible to obtain information on the type of this motion. The results were found between the case of free rotation and of rotational diffusion (compare Eqs. 164 and 165); this has been taken into account by the introduction of a *randomizing* (damping) mechanism into the free rotational correlation functions. Furthermore, from the elastic intensity as a function of κ the ratio of the time during which the rotation is nearly free, and the time when it is bound, τ_1/τ_0, has been calculated by making use of the formalism (167). The results have indicated that τ_1 is not negligible compared to τ_0.[18]

8. Molecular Liquids

8.1 Remarks on Theory

In describing the scattering on protons in simple molecular liquids we can follow the ideas which have been developed extensively for molecular crystals in Section 7.2. They were based on the simplifying (and not always justified) assumption that translational and rotational motions are uncorrelated. Therefore we factorize the intermediate scattering function as in Eq. (153), writing

$$I_s = I_s^{(r)} I_s^{(t)}. \tag{170}$$

Here, for the liquid, $I_s^{(t)}$ contains the diffusive motion of the molecular center of mass [128].

---

[18] Further experiments on the same material [137] above T_c have been interpreted in a completely different concept, namely in terms of a jump model as in Section 7.1, neglecting the time τ_1 for the rotations between discrete quasi-equilibrium positions. The resulting jump rate $1/\tau$ was of the order of 10^{11} sec^{-1}.

If the rates of rotational and translational motions differ sufficiently, the quasielastic spectrum again reveals two components (see Fig. 4, Section 2): One of them, being a delta-function for the solid, has a width $\Gamma^{(t)} = 2\hbar\kappa^2 D$, as determined by the self-diffusion constant D of the molecule. The width of the other contribution, $\Gamma^{(r)}$, is due to some kind of molecular rotation. It is $\Gamma^{(r)} \simeq 2\hbar D_r$ for rotational diffusion; for a nearly free rotation one would get a spectrum according to Eq. (160) with a side-maxima separation (or half-width for a damped rotation) of $\Gamma^{(r)} \simeq \hbar(k_B T/I)^{1/2}$. As a consequence, a separation of the rotational and the translational components would be possible if

$$\kappa^2 D < D_r; \quad \text{or} \quad < (k_B T/I)^{1/2} \tag{171}$$

provided that the experimental resolution is sufficiently good.

The criterium (171) is difficult to fulfill for liquids of normal viscosity and/or for molecules with large moments of inertia I. A reduction of κ would not improve the situation because this correspondingly reduces the intensity of the rotational component (see Eq. (158)).

A model for a molecular liquid which is more general than the one underlying Eq. (170) has been worked out by Larsson [138]; it is based on the formalism of Singwi and Sjölander as described in Section 5: The translational and rotational motion of a proton in a molecule is supposed to consist of two components; one of them, concerning the *translation* of the whole molecule, is assumed to be a vibrational state during an average time τ_0', and a diffusive state during a time τ_1'. The macroscopic diffusion constant is then given by $D = D_e \tau_1'/(\tau_0' + \tau_1')$ where D_e is the (unobservable) diffusion constant during τ_1'. Independently of this translation the scattering proton performs *rotational* motions around the center of mass of the molecule; this consists of an alternation between librations during a time τ_0, and some rotation during time τ_1 which leads to a change of the orientation [19]. The rotational part can be described with the concepts of Section 7.2.

The resulting incoherent scattering law is then of a similar mathematical structure as Eq. (167), after proper redefinition of the functions a and b. They now contain the quantities $\tau_0', \tau_1', \tau_0, \tau_1, D_e$, and the parameters characterizing the vibrational and the librational motions.

8.2 Experiments

Liquid methane has been investigated by neutron scattering with low [140] an with rather high incident energies and momentum transfers [141]. The results have been compared with theory [124, 128] using the facto-

---

[19] See the footnote at the end of Section 7.2.

rization of Eq. (170). The translation has been treated in terms of a delayed diffusion model as in Section 4.2 with the experimentally known value of D, and with an estimated "typical oscillation frequency" of the molecules in the liquid. The rotational part has been treated by means of (155) or (166), using the orientational correlation functions $F_1(t)$ and $F_2(t)$ only. These have been taken from infrared and Raman scattering experiments, respectively (see Fig. 36). Terms with $l > 2$ in the angular momentum expansion were small. The agreement obtained between these calculations and the neutron data turned out to be fairly good.

This consistency between two completely different methods demonstrates the reliability of the underlying physical concepts. Therefore, one can hope that neutron scattering experiments with higher accuracy could give information on molecular rotations, especially on the leading term $\langle P_1 [\cos \beta(t)] \rangle$ and, with less accuracy, also on the higher correlation functions. It should be pointed out that for methane the interpretation of the results is particularly transparent because the interaction between the molecules is relatively weak; furthermore, the CH_4-molecule is nearly spherical, and rotational-translational coupling is expected to be small.

Further work on *liquified or dense gases* will only be mentioned, namely on liquid H_2 and D_2 [142], on H_2 and CH_4 dissolved in liquid noble gases [143] and on NH_4 and H_2O vapors [144, 145]. We emphasize that, in general, the study of gases at moderate density is particularly interesting (although difficult experimentally) because the rotational and translational motions can be treated more from first principles than it is possible for liquids.

As an example for investigations on a strongly associated liquid we report experiments on the quasielastic width of *glycerol*. The first studies have been performed with an energy resolution of about $2 \cdot 10^{-4}$ eV by Larsson et al. [139, 146]. The slope of Γ vs. κ^2 for small κ has been interpreted in terms of an effective diffusion constant as defined from the measured half-width $\Gamma_{eff} = 2\hbar\kappa^2 D_{eff}$; this includes translational as well as diffusional contributions *(Fig. 29)*. For large temperatures the viscosity is low, and D_{eff} is closely proportional to the diffusion constant D which has been estimated by means of Eyring's formula $D = k_B T / M\eta$ from viscosity data $\eta(T)$. At lower temperatures, however, D_{eff} turned out to be considerably higher than D. This can be understood as follows: At low temperatures, where the diffusion of the molecule as a whole is very slow, the motion of the protons and the quasielastic width is determined by rotational jumps of the OH-groups. The motion of such a group needs a smaller number of bonds broken than the motion of the whole molecule; as a consequence, the slope of $\ln D_{eff}$ vs. $1/T$ and,

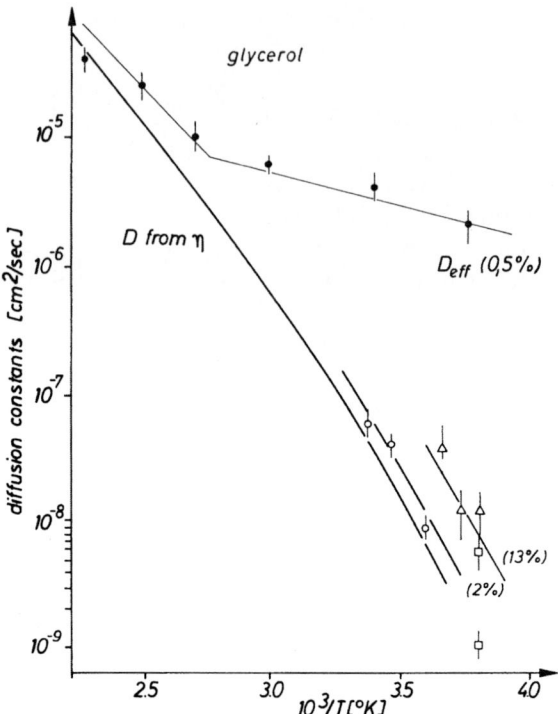

Fig. 29. Diffusive motions in glycerol. Solid circles: D_{eff} of protons in glycerol as defined by the width of quasielastic scattering at small κ, namely $\Gamma = 2\,\hbar k^2 D_{eff}$, as a function of temperature (resolution $\sim 2 \cdot 10^{-4}$ eV). From Larsson and Dahlborg [146]. Open circles and triangles: Measurements with the 90°-spectrometer. From Birr [147] (see Fig. 30). Thin lines are drawn as guide-lines for the eye. Squares: Mössbauer effect with cobalt chloride; from Craig and Sutin [148]. Solid line: Self-diffusion constant calculated from viscosity η. Water content of glycerol is indicated in brackets

therefore, the apparent activation energy is several times smaller than in the region of higher temperatures (Fig. 29).

More recently, Birr [147] has performed measurements on glycerol by means of the backscattering spectrometer with a 300 times better energy resolution of about $7 \cdot 10^{-7}$ eV (Section 3.2). *Fig. 30* shows typical quasielastic spectra. The reader should notice the low count rates taken with this instrument, as a consequence of the extremely high resolution. The effective diffusion constant D_{eff}, as determined from the half-width in these measurements, is also included in Fig. 29. Here the activation energy of D_{eff} and of D (from η) agrees in the whole region of temperatures.

These results can be understood as follows (see Fig. 4). The quasi-elastic spectrum consists of a relatively sharp line, the width of which is essentially determined by the translational motion of the whole molecule (i.e. by D) plus a ~ 100 times broader quasielastic part; the latter might result from some kind of rotational motion. As a consequence, in the very high resolution experiment this quasielastic broad line will appear as a flat background which cannot be observed at all; only the sharp line is being seen. On the other hand, it can be suspected that in the experiments with relatively poor resolution this broad contribution merges

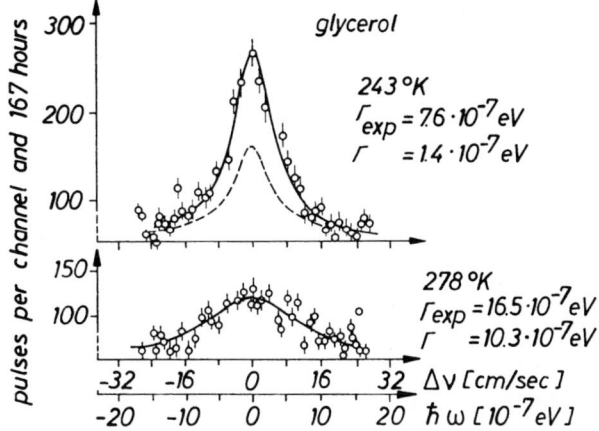

Fig. 30. Typical quasielastic spectra of glycerol measured by means of the back-scattering spectrometer. Dashed: Resolution curve (width $7 \cdot 10^{-7}$ eV) as determined from purely elastic scattering on a vanadium sample. Γ_{exp} experimental width; Γ resolution-corrected quasielastic width (From Birr [147])

with the sharp line [138]; the resulting effective width will be somewhere between the width of the broad and the sharp line. The supposition that the high resolution experiment does not observe the whole spectrum is supported by a determination of the absolute line intensity by means of a vanadium standard scatterer. It has been found that the line includes not more than 50–70 % of the intensity to be expected for a *bound* proton. Only part of this deficiency can be attributed to the Debye-Waller factor.

Besides these extensive studies on glycerol there exists a great number of investigations on other liquids; the references we give may be considered only as examples taken from a wide field of chemical applications of the method [149–152, 138, 139]. In this connection, we especially point out the possibility to separate different kinds of motions of a molecule; this is feasible by replacing certain hydrogens by atoms with smaller cross sections [40].

9. Polymeres and other Complicated Systems

Here we include a number of topics which have in common a greater degree of complexity than those which have been treated in the previous sections. Not much work has been devoted to this field and our aim is mainly to draw attention to possible applications of the method, in particular in polymeric chemistry and biology.

9.1 Polymers

The motion of an atom in a polymeric chain can be described by a superposition of different components which either contribute to the quasielastic or to the inelastic spectrum of the scattered neutrons.

(i) *Oscillations.* Collective waves propagating along the chains ("one-dimensional phonons") as well as localized vibrations and librations of side-groups give rise to an inelastic spectrum. They also determine the intensity of quasielastic scattering via the Debye-Waller factor (see for instance [153]).

(ii) *Side-group Motions.* Side-groups of the polymeric chains, as CH_3-groups, are able to perform rotational jumps around their bond axes. Such motions can be treated in terms of the models discussed in Section 7. At elevated temperatures the corresponding jump rates can be as high as 10^8 to $10^{12}\ sec^{-1}$ so that they are visible in the neutron spectrum.

(iii) *Motion of Individual Chain Links.* In a polymeric solid the chains do not form regular structures. The disorder is made up by faults with respect to the mutual orientation of adjacent chain segments. Part of these faults are not static; for instance, they are able to perform rotational motions or to migrate along the chains. As an example, *Fig. 31* demon-

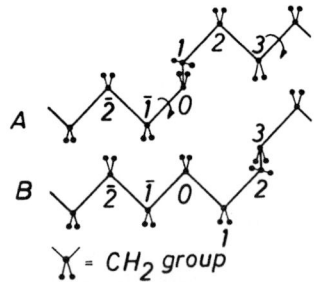

Fig. 31. Model for a kink jump process in polyethylene. Conformation A: Chain sections $\bar{2}$, $\bar{1}$, 0 within, and 1, 2, 3 above paper plane. Bond 0–1 points out of the plane. Conformation B: $\bar{2}$, $\bar{1}$, 0, 1, 2 within, 3 above paper plane. The kink has been displaced along the chain by simultaneous rotation of atoms 1 and 2 around a C–C bond axis. The other atoms remain fixed (see [154])

strates how the migration of a "kink" could be imagined. It consists of a simultaneous and jumpwise rotation and displacement of two CH_2-groups [154], whereby the other groups, i.e. the remaining parts of the chain, stay fixed.

Scattering on these protons could be described by means of the theory in Section 7.2, using a jump model with two quasi-rest positions. Consequently, the resulting scattering law will contain a quasielastic part; its width would be of the order of $\hbar/\tau$ (Eq. (152)), where τ is the average time between jumps which an individual proton undergoes[20]. In addition, the spectrum contains a line at $\hbar\omega = 0$ which is sharp if the chain as a whole is fixed. The area of the sharp elastic line depends on the jump geometry. If a certain fraction of the chain segments is completely immobile, the intensity of the sharp line is correspondingly larger. Therefore, a determination of the elastic and the quasielastic intensity might give information on this fraction.

(iv) *Diffusive Motions.* Now we consider dense polymeres having finite viscosity, or dilute polymeres. In such cases, the motions described before are superimposed with a slow diffusion of the chain segments. This leads to a broadening of the line which is sharp for the solid polymere.

At very large times, the mean square displacement of any atom of the chain should approach normal diffusive behaviour; here, the mean square displacement is given by

$$\langle r^2(t) \rangle = 6Dt + \text{const}. \tag{172}$$

Large times means $t \gg R^2/D$, where D is the self-diffusion constant of the center of gravity, and R is the radius of gyration of the chain. We expect that the constant in (172) is of the order of R^2. This motion would only be observable at extremely small κ-values $(\kappa < 1/R)$; the corresponding quasielastic width would be $\Gamma = 2\hbar\kappa^2 D$.

For intermediate times, the situation is more complicated. The links of the chain are performing a slow motion which differs from the diffusion discussed in terms of the Langevin equation. Making the simplifying assumption that there is no interaction between different chain links, this motion can be treated in the frame of the theory of Rouse [155]; the pertinent result in this case is [156] that the mean square displacement is given by

$$\langle r^2(t) \rangle = R(12Dt/\pi)^{1/2}. \tag{173a}$$

Eq. (173) with (37) leads to a quasielastic scattering distribution with an exceptional κ-dependence of its half-width, namely

$$\Gamma \simeq 6 \cdot 10^{-2} \hbar D R^2 \kappa^4. \tag{173b}$$

---

[20] Obviously, $1/\tau$ is not necessarily identical with the jump rate of the kink.

This holds for a region where κ is larger than $1/R$, and sufficiently small compared to the reciprocal length of the chain segments.

Actually, interactions between different segments of a chain cannot be neglected [157]. One kind of such an interaction is due to the velocity field which the segments of the chain induce in the surrounding solvent and which acts upon the other segments[21]. Under this assumption, one gets a power law for the mean square displacement [159], which is

$$\langle r^2(t) \rangle \sim t^{2/3} \quad \text{and} \quad \Gamma \sim \kappa^3 .$$

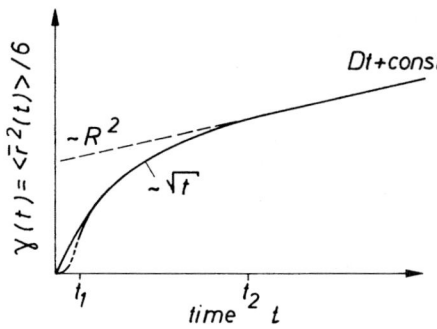

Fig. 32. Qualitative sketch of mean square displacement of an atom in a dilute polymere. $t < t_1$: jump motions; $t < t_2$: diffusion according to "internal" motions of the chain (see Section 9.1). For $t > t_2$ a diffusive behaviour of the center of gravity is approached. R radius of gyration of the chain

For dilute polymeres, this motion is superimposed with a slow *rotational diffusion* of the whole polymeric coil around its center of gravity. Such kind of motion could be treated according to the formalism given in Section 7.2.

In *Fig. 32* the qualitative features of the mean square displacement as a function of time are shown. It should be noticed that in the region where t is of the order of the mean jump time τ, $G_s(r, t)$ deviates strongly from a Gaussian (dashed part of the curve in Fig. 32). The corresponding behaviour of the quasielastic spectrum can be summarized as follows. The spectrum is composite: There will be a quasielastic part around $\hbar\omega = 0$ with a width determined by the jump motions (see ii and iii). The intensity of this part vanishes if κ goes to zero. In addition, there is a sharper component which is determined by the complicated diffusive motions as described under iv.

Numerical estimates by means of (173) lead to halfwidths smaller than the best resolutions obtainable so far, except for very short chains and for elevated temperatures. It might be feasible, however, to investigate

---

[21] For a treatment of semi-dilute polymeres see [158].

the intermediate region where the diffusion occurs over a relatively small number of steps. This region has not yet been treated theoretically.

An inherent complication for polymeres with sidegroups is the coexistence of jumps of the side-groups and of the chain segments; both produce quasielastic lines. In certain cases, this difficulty might be overcome if side-group protons are being replaced by atoms with a smaller scattering power.

Experimentally, quasielastic scattering has been studied on polymeric *liquid selenium* [160] as a function of temperature and of the chain length; the latter can be changed by adding different amounts of iodine. The quasielastic width has been found to be much greater than it should be according to the self-diffusion constant of Se in the melt. Consequently, the spectra must be interpreted in terms of a jump model similar to that discussed under (iii). A separation of the elastic and the quasielastic components has not been possible. Therefore, the measured half-width has been compared with the effective width which one obtains from a calculated composite spectrum, broadened by the instrumental resolution. It turned out that the effective width, and therefore the jump rate, did not depend on the chain length within experimental error, in contrast to the viscosity. The jump rate $1/\tau$ calculated from the spectra varied between 10^{11} and 10^{12} sec^{-1}; the activation energy of $1/\tau$ was 8 kcal/mol. This is much smaller than the apparent activation energy of the viscosity which is between 12 and 18 kcal/mol. These observations confirm the idea that the quasielastic width is in fact determined by a "local" step of the diffusive process. In this case, it is difficult to find the geometry of the jump process because in liquid selenium a considerable fraction of the atoms is bound in Se_8 rings rather than in chains.

There is also evidence for the existence of polymeric chains in liquid *hydrogen fluoride*. Neutron scattering experiments [161] have shown that the quasielastic intensity drops much steeper than could be expected on the basis of a Debye-Waller factor (see Sect. 7.1). This suggests an interpretation in terms of a random rotation around the chain axis of the protons, sitting on the segments of the zigzag-shaped molecules. Also in this case it has not been possible to separate the purely elastic term $l = 0$ in Eq. (158).

A problem of broader interest is the investigation of polymeric hydro-carbons. Experiments in this direction are being prepared in several laboratories.

At the end of this section we comment briefly on a relatively new field of quasielastic scattering, namely on *liquid crystals*. Here, again the quasielastic spectrum is determined by a composition of translational and rotational motion. First experiments have been reported on nematic *p*-azoxyanisol under an external electric field [162] which orients the molecular axis parallel to the field vector. A considerable anisotropy of the quasielastic width has been found. From this it has been tried to evaluate the anisotropy of the diffusive motion. An extensive study of various liquid crystals in their different phases has been performed by Blinc *et al.* [163]. The "effective diffusion constant" has been compared with the true diffusion constant, as determined by the gradient spin echo method. The interpretation of the results is difficult since the effective diffusion constant depends on translational diffusion as well as on rotational motions; the results of this comparison will not be discussed here.

9.2 Different Kinds of Water

Depending on the kind of environment around a H_2O molecule, condensed water can exist in various configurations; they differ from each other by structure, vibrational spectrum, and mobility. Neutron scattering has widely been used to study these effects.

A number of extensive experiments has been reported on *aqueous solutions* of salts and alcohols as a function of concentration and temperature [88, 164, 165], on water and other liquids in molecular sieves [166], on aqueous gels and glasses [167, 168] and on adsorbed gases [169], a research field of particular importance for chemical catalysis. *Fig. 33* gives an example of such measurements on H_2O in silica gels at different concentrations. In most of these cases the quasielastic width has been

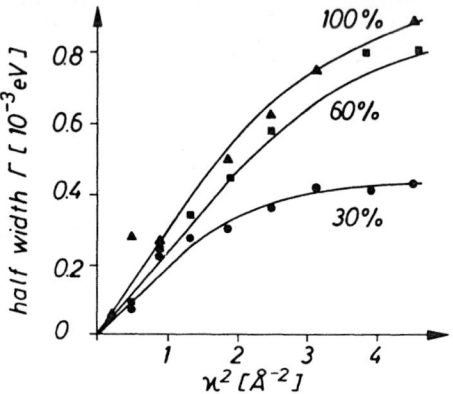

Fig. 33. Width of quasielastic scattering for H_2O in silica gels as a function of κ, indicating the change in the self-diffusion constant and the jump rate for various H_2O/SiO_2-ratios (mass per cent) (From Anderson and White [168])

evaluated in terms of the step model discussed in Section 5. From this model the self-diffusion constant of the H_2O molecule and its rest time τ in a hydrogen bond have been evaluated. The application of this step diffusion model is adequate to aqueous solutions. For H_2O in gels, however, it has to be considered that the quasielastic width might be dominated by rotational jumps whereas the translational diffusion (jumps from cage to cage) is slow. In such a case, the spectrum will again be of the composite type and only an effective width of the spectrum might be observed experimentally.

Another kind of "bound water" is found, to a large extent, in *biological matter*, like living cells or muscles. Nuclear magnetic resonance and other evidence have demonstrated that the mobility of these H_2O molecules, or of a certain fraction of them, is smaller than the mobility in the liquid. This suggests some form of ice-like coordination of the H_2O close to the surface of the biological molecules [170]. Quasielastic scattering might be able to give supplementary information on the diffusive behaviour of such H_2O molecules. Furthermore, the fraction of the molecules having such a reduced mobility could be evaluated by determining the intensity of the sharp and the broad contribution of the quasielastic spectrum (if this exists), as a function of κ.

10. Effects of Coherent Scattering

In the previous sections we have mainly been dealing with incoherent quasielastic scattering on hydrogeneous substances; here the scattering law depends on the individual motion of the scattering protons. For substances where the coherent contribution is dominating, the scattering is determined by the motion of *pairs* of atoms relative to each other, and $S_{coh}(\kappa, \omega)$ depends on the correlations between the motion of different atoms. In discussing the subject, two domains have to be considered.

(i) the *hydrodynamic region*, where slow and long-range collective motions of the particles play the major role. This behaviour is to be observed at very small values of κ and ω ($\kappa \lesssim 10^{-3} \text{Å}^{-1}, \omega < 10^{10} \text{sec}^{-1}$) which is the typical domain of Laser spectroscopy (Fig. 6). This field is theoretically well established in terms of thermodynamics and continuum mechanics (see for instance [171–173]); the corresponding scattering law can be directly applied to coherent neutron scattering [174]. Because of intensity reasons the use of neutrons in this region is extremely difficult. Therefore, we will deal with this subject only briefly.

(ii) As soon as $1/\kappa$ becomes comparable with the interatomic distances and as the corresponding frequencies become high, the hydrodynamic theory fails. In this "*quasi-crystalline region*" the system can no longer be considered as a continuum obeying thermodynamical equations. We restrict ourselves to a brief discussion of quasielastic coherent scattering in this region for a simple liquid; furthermore, we add a remark on collective rotations in molecular crystals.

10.1 Hydrodynamic Description

As has been pointed out earlier, the scattering process performs a Fourier analysis of the atomic motions in space and time. We assume here that the period $2\pi/\omega$ of the Fourier component under investigation is small compared to the typical relaxation times of the liquid; furthermore, we consider κ-values small compared to the reciprocal interatomic distances, $1/d$. To calculate $S_{coh}(\kappa, \omega)$ we now consider the correlation function in the form (44) of the density auto-correlation function, $G(r, t) = \langle \varrho(0, 0) \varrho(r, t) \rangle / \bar{\varrho}$.

Because we have $\kappa \ll 1/d$, we are allowed to neglect the atomistic structure of the liquid as reflected by the δ-functions in the microscopic density $\varrho(r, t)$. Consequently we can treat the liquid like a classical continuum with a density $\varrho(r, t)$, being a slowly fluctuating function of space and time. The behaviour of $\varrho(r, t)$ and, from this, the auto-correlation function can be found by solving the hydrodynamic equations [171–174]. The underlying assumption is that the thermodynamic variables, for instance pressure, density and temperature, are interconnected by the macroscopic (equilibrium) equation of state $F(\bar{\varrho}, p, T) = 0$; it is furthermore assumed that the dissipative properties of the liquid are described by linear macroscopic transport equations, for instance for the heat current

$$J = -\lambda_T \, \mathrm{grad} \, T \tag{174}$$

where λ_T is the thermal conductivity.

It then turns out that the density fluctuations $\varrho(r, t)$ can be decomposed into two independent contributions, namely propagating contributions having the character of *longitudinal adiabatic sound waves* with the velocity c_{ad}, and a contribution originating from fluctuations of the local temperature; normally, the latter are of the *diffusive type*. Each sound wave imposes onto the liquid a periodical "lattice" of density fluctuations which is propagating through the liquid. As a consequence, the neutron wave suffers inelastic diffraction obeying the Bragg law for this "lattice", $\kappa = q$ (q is the wave vector of the scattering sound wave). This leads to the well-known Brillouin doublet with lines at $\omega = \pm c_{\mathrm{ad}} q$. The width of these lines is determined by the heat conductivity and by the viscosity of the liquid. The non-propagating fluctuations create a quasielastic peak around $\hbar\omega = 0$ with a half-width

$$\Gamma = 2\hbar\kappa^2 D_{\mathrm{th}} \tag{175}$$

where $D_{\mathrm{th}} = \lambda_T / c_p \bar{\varrho}$ is the heat diffusion constant.

The total intensity of the resulting triplet is found to be

$$S(\kappa) = k_B T (\partial \bar{\varrho} / \partial p)_{T'} \quad \text{for} \quad \kappa \to 0. \tag{176}$$

From thermodynamic arguments one further finds the intensity of the doublet $2I_1$ relative to the intensity of the central line I_0, namely

$$2I_1 / (2I_1 + I_0) = (\partial \bar{\varrho} / \partial p)_S / (\partial \bar{\varrho} / \partial p)_T. \tag{177}$$

Here $(\partial \bar{\varrho} / \partial p)_T$ and $(\partial \bar{\varrho} / \partial p)_S$ are the isothermal and the adiabatic compressibilities.

10.2 Influence of the Liquid Structure

For scattering vectors κ being of the same order of magnitude as the reciprocal interatomic distances, the description of the liquid in terms of a smoothly varying density $\varrho(r, t)$ fails. As a consequence, other theoretical concepts are needed. An approach which has been pursued by several authors is to relate the time-dependent pair correlation function $G_d(r, t)$ with the structural properties of a liquid, described by $g(r)$, and with its dynamical behaviour by means of $G_s(r, t)$.

As a first approach we introduce the following concept *(Fig. 34)*: Assume that an atom A is for $t = 0$ at the origin. Then $g(r')\,dr'$ is the

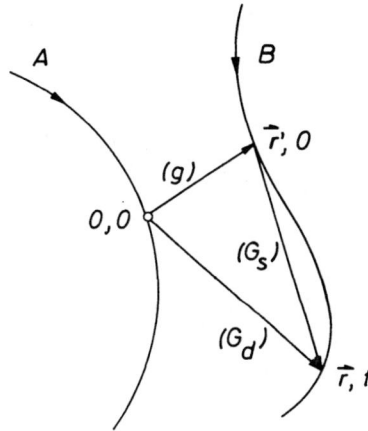

Fig. 34. Paths of two distinct particles for the calculation of $G(r, t)$ from the static pair correlation function $g(r)$ and the self-correlation function $G_s(r, t)$ in the convolution approximation (Eq. (180))

probability of finding simultaneously an other atom B in $d r'$ at r'. Now we introduce a function $H(r, r', t)$ which describes the probability that *atom B migrates from r' to r in time t conditioned by the fact that atom A was at $r = 0$ for $t = 0$.* Consequently, $g(r')\,dr'\,H(r, r't)$ is the probability to find an atom B in $d r'$ at r' for $t = 0$, migrating to r in time t. The total probability of finding atom B at r for time t wherever it was for $t = 0$, then is

$$G_d(r, t) = \int g(r')\,H(r, r', t)\,dr' .\tag{178}$$

As an approximation we now replace the conditioned probability function H by the self-correlation function; this means

$$H(r, r', t) \simeq G_s(r - r', t)\tag{179}$$

where G_s is defined as the probability that an atom migrates from r' to r in time t, *without* taking into account the position of any other atom. Recalling that $G(v, t)$ is the sum of G_s and G_d, the so-called *convolution approximation* is obtained (Vineyard [16]), namely

$$G^{(\text{conv})}(v, t) = G_s(v, t) + \int g(v')\,G_s(v - v', t)\,dv' .\tag{180}$$

In Fourier space, Eq. (180) is equivalent to a replacement of

$$I(\kappa, t) = N^{-1} \sum_i \sum_j \langle \exp\{i\kappa[r_j(t) - r_i(0)]\}\rangle$$

$$\equiv N^{-1} \sum_i \sum_j \langle \exp\{i\kappa[r_j(t) - r_j(0)]\} \exp\{i\kappa[r_j(0) - r_i(0)]\}\rangle \tag{181}$$

by the factorization

$$I^{(conv)}(\kappa, t) = N^{-1} \sum_i \sum_j \langle \exp\{i\kappa[r_j(t) - r_j(0)]\}\rangle \langle \exp\{i\kappa[r_j(0) - r_i(0)]\}\rangle$$

$$\equiv I_s(\kappa, t) I(\kappa, 0). \tag{182}$$

Consequently, for the scattering law one obtains

$$S^{(conv)}(\kappa, \omega) = S(\kappa) S_{inc}(\kappa, \omega) \tag{183}$$

which means that the energy distribution for coherent scattering would be proportional to $S_{inc}(\kappa, \omega)$.

It is obvious that the approximation *fails for small distances;* here the motion of particle B is strongly correlated with particle A, at least by the fact that A and B cannot overlap. Therefore, (182) fails, if $1/\kappa$ is comparable with the interatomic distances. The negligence of correlations can directly be recognized in the space-time behaviour of $G_d(r, t)$ shown in Fig. 3, where $G_d^{(conv)}(r, t)$ smoothens out too fast for increasing time.

Unfortunately, also for $\kappa \ll 1/d$ (hydrodynamic limit) the convolution approximation turns out to be wrong; it predicts a central line with a halfwidth $2\hbar\kappa^2 D$, where D is the self-diffusion constant. Hydrodynamic theory, on the other hand, predicts the Brillouin lines with a middle peak; its width will be given by the 10–100 times larger heat diffusion constant, D_{th} (Eq. (175)): As a matter of fact, the *diffusive* motions of particles separated by large distances are uncorrelated; they are, however, superimposed by the density fluctuations discussed in Section 10.1 which are of long range.

Singwi [175, 176] has worked out a simple physical model which is able to improve the convolution approximation in the non-hydrodynamic κ-region. For large atomic pair distances, $|r_i - r_j| > R_c$, the double sum in (181) is treated according to the convolution approximation. This means that the movement of the atoms i and j is assumed to be independent. For $|r_i - r_j| < R_c$, correlations are taken into account by treating, as a rough approximation, the liquid like a harmonic crystal with phonons. The correlation distance R_c is considered as an empirical model parameter. The occurence of phonon-like excitations in liquids in the non-hydrodynamic region (large κ and ω) is a subject of considerable interest and controversy. There is some evidence from neutron inelastic scattering that such excitations exist [45], at least in liquid metals, so that the supposition of the Singwi model might be justified.

It is somewhat artificial to consider quasielastic coherent scattering without dealing with the inelastic (quasi-phonon) part of the spectrum; the problem has, in fact, to be treated as a whole. In this connection we draw the attention to the self-consistent field approach of Singwi *et al.* [178] which is rather transparent from a physical point of view, and to the treatment in terms of a generalization of hydrodynamics by Chung *et al.* [181]. A number of further theoretical investigations on the coherent scattering law in liquids has been published [179, 180, 182], where the main emphasis is on the *collective* aspects of the motion; their treatment is beyond the scope of this review.

An alternative way of calculating G_d makes use of the function $P(r, r', t)$ which describes the conditional probability that a certain particle A is at point r for time t, given that *another particle B has reached point r' at the same time t, which was at $r' = 0$ for $t = 0$.* Now we average over the probability $G_s(r', t)$ for the particle B moving from $r' = 0$ at $t = 0$ to r' at t. This leads to [47, 177]

$$G_d(r, t) = \int P(r, r', t)\, G_s(r', t)\, \mathrm{d}r' . \tag{184}$$

Here, the convolution approximation (180) would be obtained by assuming that

$$P(r, r', t) \simeq g(r - r') . \tag{185}$$

This is expected to hold for large times where the particle has forgotten its history[22]. The conditioned probability $P(r, r', t)$ can be calculated by simple physical models [47, 72], for instance by assuming that the approach of this function to its stationary value (185) follows a relaxation equation

$$\partial P(r, r', t)/\partial t = -(1/\tau_p)\,[P(r, r', t) - g(r - r')] \tag{186}$$

with a certain decay time τ_p. This leads to the following expression

$$G_d(r, t) = G_d^{(\mathrm{conv})}(r, \hat{t}) + \exp\{-t/\tau_p\}\,[g(r) - G_d^{(\mathrm{conv})}(r, \hat{t})] \tag{187}$$

where $\hat{t}(t)$ is defined similarly to the "delayed time variable" introduced by Rahman [14].

It has to be emphasized that the *experiments* are strongly contradicting the factorization (183) predicted by the convolution approximation. The striking effect, observed experimentally, is an oscillatory behaviour of the quasielastic width; it shows a minimum at approximately that κ-value where the structure factor has its maximum [77, 43]. This "*line narrowing*" has been predicted, for the first time, directly by means of the sum rules [23]; as has been pointed out earlier, the minima of the function $\langle\omega^4\rangle_{\mathrm{coh}}/\langle\omega^2\rangle_{\mathrm{coh}}^2$ (Section 2.4) give an indication for a contraction of the energy distribution (see also [186]). (Fig. 5). This

---

[22] In contrast to the convolution approximation, G_s has been introduced rigorously into (184), and P is approximated. (The variables in Eq. (184) do not coincide with those of Fig. 34.)

line narrowing is reproduced by the quasi-crystalline model [175] mentioned before, in rough agreement with experimental data. Existing experiments are also reproduced by the theory of Singwi *et al.* [178] developed more recently *(Fig. 35)*. Andriesse [183] has systematically determined the difference between the experimental scattering law for argon, and the results due to the convolution approximation (the self-

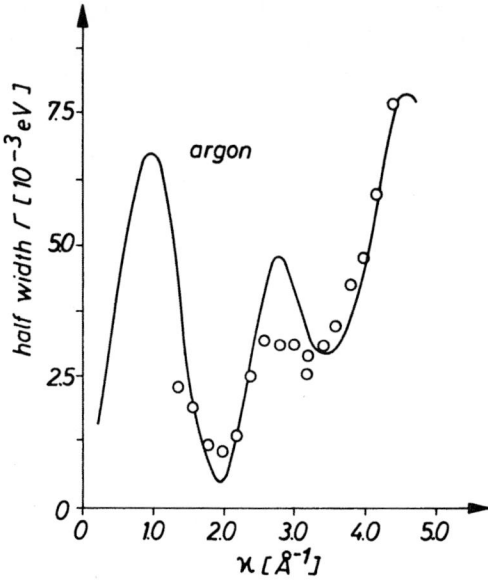

Fig. 35. The effect of "line narrowing" of the quasielastic width $\Gamma(\kappa)$ for liquid A^{36}; it appears in the vicinity of the peaks of the structure factor. Preliminary experimental points from Sköld *et al.* [43]. Solid line: Theory of Singwi *et al.* [178]

part has been calculated as in Section 4). So far, coherent scattering investigations have not been able to give essentially new physical insight into the atomic motions in liquids. However, this field is of great theoretical interest and progress is to be expected in future work.

It should be mentioned that, according to computer experiments, a narrowing of the quasielastic line also occurs for *incoherent* scattering if $1/\kappa$ is comparable with the interatomic distances [17, 181]. It seems that this is inherently connected with the appearance of deviations from a Gaussian-shaped selfcorrelation function.

Finally, we want to mention briefly a completely different concept for the interpretation of line narrowing [200]. It introduces a diffusive mode at wave vectors close to the first peak of the structure factor;

the relaxation time of this mode is supposed to increase close to the melting temperature.

So far, to our knowledge, no investigations exist concerning coherent quasielastic scattering on non-oscillatory *cooperative, rotations* in molecular crystals[23] and liquids. From incoherent and coherent neutron scattering one could obtain information concerning the different decay times of the correlation functions for individual and for collective rotations. Furthermore, coherent scattering would yield the range of the correlations. Unfortunately, the pertinent spectra are difficult to interpret: They contain contributions due to the *self*-motion of the atoms in the molecule as well as interference contributions from *intra* and from *inter*-molecular scattering.

The observability of such cooperative effects has been demonstrated for a special example [127] comparing incoherent neutron and (coherent) light scattering experiments on plastic cyclohexane. By measurements of structure factors [185] it has been shown that at least static orientational correlations exist for simple molecular liquids.

A particular situation occurs close to the *critical point* of an orientational phase transition. Here the average orientation can be treated in terms of an order parameter. Large orientational clusters could appear with molecules in preferential orientations; these clusters might undergo slow fluctuations as a function of time. If only two orientations exist and if the coupling of the rotating molecule with other degrees of freedom of the lattice is negligible, the dynamical behaviour of the fluctuations can be described like heat or spin diffusion [174]. As a consequence, close to T_c the quasielastic width of coherent scattering on these fluctuations would behave according to equation (175) where D_{th} plays the role of a diffusion constant for the molecular orientations.

11. Quasielastic Scattering and other Methods

A comparison between the different methods and a judgement of their merits and disadvantages is difficult. Therefore, we restrict ourselves to a discussion of a few aspects which quasielastic scattering has *in common* with other methods. Such considerations might help to enable the experimentalist to decide if the application of neutron spectroscopy is useful, or if similar results can be gained from other and less expensive methods. In such considerations one has to realize that neutron spectroscopy has reached the state of a routine method, like *nmr* or infrared spectroscopy, so that the application of neutron scattering is no more an interesting subject in itself.

11.1 Various Kinds of Scattering Experiments

It has already been mentioned that a number of interactions can be described by the same, or by similar theoretical concepts as those used for quasielastic scattering; this fact has its origin mainly in the applicability of first order perturbation theory. As has been pointed out, X-ray diffraction is able to determine the structure factor

---

[23] On the other hand, *oscillatory* collective rotations (librons) have been investigated in many instances [184].

$S(\kappa) = \int S_{\mathrm{coh}}(\kappa, \omega) \, d\omega$; light scattering on weak collective density fluctuations in the hydrodynamic region is proportional to the Fourier transform of the density auto-correlation function (44).

An important subject is inelastic light scattering on rotational motions of *individual* molecules in solids and liquids [187, 188]. In *infrared spectroscopy* the intensity distribution of the scattered light depends on the time-dependent fluctuation of the dipole moment vector $\boldsymbol{u}(t)$ of the scattering molecule, namely

$$I(\omega) \sim \int_{-\infty}^{+\infty} \exp\{-i\omega t\} \, \langle \boldsymbol{u}(0) \, \boldsymbol{u}(t) \rangle \, dt \, . \tag{188}$$

ω is the frequency displacement of the scattered light from the vibrational band center. The dipolar correlation function $\langle \boldsymbol{u}(0) \, \boldsymbol{u}(t) \rangle$ is defined as in Eqs. (156) or (157).

Raman scattering concerns the rotational motion of the polarizibility ellipsoid. For the special case of a symmetric vibration in a symmetric-top or linear molecule one obtains a similar relation [187]

$$I(\omega) \sim \int_{-\infty}^{+\infty} \exp\{-i\omega t\} \, \langle P_2[\boldsymbol{u}(0) \, \boldsymbol{u}(t)] \rangle \, dt \tag{189}$$

where $P_2(x) = (3x^2 - 1)/2$. Here, $\boldsymbol{u}(t)$ is a time-dependent vector along the symmetry axis of the molecule (see *Table 4*).

On the other hand it has been shown in Section 7 that the *incoherent neutron scattering* law can be developed into a series of terms

$$j_l^2(\kappa d) \, \langle P_l[\boldsymbol{u}(0) \, \boldsymbol{u}(t)] \rangle \tag{190}$$

where $\boldsymbol{u}(t)$ is now the position vector of the scattering proton in the rotating molecule.

Consequently, Fourier inversion of light spectra allows a direct determination of the time-dependent orientational correlation functions

Table 4. Time-dependent correlation functions from various experimental methods
(see Ref. [19])

$S_{\mathrm{inc}}(\kappa, \omega)$ from neutron scattering on liquids	$\langle v(0) \, v(t) \rangle$
$S_{\mathrm{inc}}(\kappa, \omega)$ from neutron scattering on rotating molecules	Superposition of functions $\langle P_l[\boldsymbol{u}(0) \, \boldsymbol{u}(t)] \rangle, \quad l = 1, 2, \ldots$
Infrared obsorption; vibration-rotation spectrum $I_{\mathrm{IR}}(\omega)$	$\langle P_1[\boldsymbol{u}(0) \, \boldsymbol{u}(t)] \rangle$
Raman-scattering; vibration-rotation spectrum; $1/T_1$ from nuclear magnetic resonance	$\langle P_2[\boldsymbol{u}(0) \, \boldsymbol{u}(t)] \rangle$

$\langle P_1 \rangle$ and $\langle P_2 \rangle$, whereas neutron scattering gives a superposition of such terms. Practically, only the first two or three of them might be obtained. Orientational correlation functions calculated by Fourier inversion of optical spectra are shown in *Fig. 36*, together with theoretical results (see also Section 8.2). It should be emphasized that in liquids also the translational diffusive motion of the molecules strongly influences the neutron spectra; in general, this is not the case for light scattering.

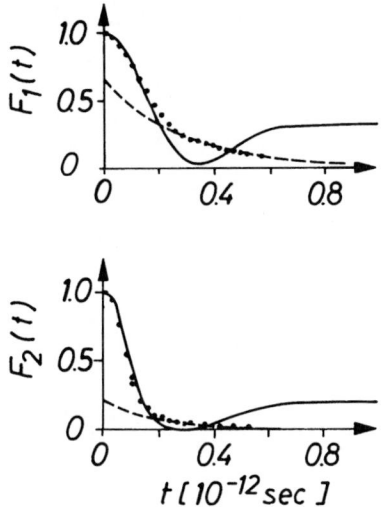

Fig. 36. Typical rotational correlation functions $F_l(t) = \langle P_l[\cos\beta(t)] \rangle$ $(l = 1, 2)$ in liquid methane at 98° K for classical free rotations (solid line; see Section 7.2). Points from infrared $(l = 1)$ and Raman scattering data $(l = 2)$. The dashed lines correspond to an exponential decay according to $\exp\{-D_l t - C_l\}$ where D_l, C_l are fitted parameters. Gordon [187]

According to Eq. (188) the intensity $I(\omega)$ for a *rotational-vibrational* combination band is connected with the *auto*-correlation function of the motion of individual molecules; the underlying assumption has been that vibrational transitions of different molecules are uncorrelated [128, 187]. In general, this assumption is justified. On the other hand, in *pure rotational spectra*[24] correlations between the rotations of *different* molecules contribute to the intensity. Consequently, the spectrum depends also on correlation functions for pairs of molecules, for instance $\langle u_i(0) u_j(t) \rangle$ (see the end of Section 10.2). As a consequence, pure rotational and rotational-vibrational spectra are, in certain respects, related to coherent and incoherent neutron scattering, respectively.

---

[24] Here, $I(\omega)$ is centered around the incident frequency.

11.2 Relaxation Methods

In discussing *nuclear magnetic resonance* we restrict ourselves to the
longitudinal *spin-lattice relaxation* rate due to magnetic dipolar interactions in a strong magnetic field H_0. This quantity, $1/T_1$, determines
the rate according to which the nuclear spin system returns to thermal
equilibrium. One can show that $1/T_1$ is proportional to the transition
rate between the energy states of the precessing spin of the reference
nucleus; the transitions are induced by time-dependent fluctuations
of the internal magnetic field $H(t)$ which is superimposed on the applied
field H_0.

The fluctuating field $H(t)$ is produced by the random rotational
and/or translational motion which the reference nucleus performs
relative to the neighbouring nuclei. As can be shown, the transition
rate is then proportional to the Fourier component of $H(t)$ at the
transition (Larmor) frequency $\omega = \gamma_I H_0$ and at its first harmonic 2ω.
Here, γ_I is the gyromagnetic ratio of the precessing nucleus having a
spin I. Explicitly, for a single pair of identical nuclei (i, j) which are
connected by a time-dependent space vector $r_{ij}(t)$ one obtains [189, 190]

$$1/T_1 = (3/2)\,\gamma_I^4 \hbar^2 I(I+1)\left[\tilde{J}_1(\omega) + \tilde{J}_2(2\omega)\right]. \tag{191}$$

The functions $\tilde{J}$ are the Fourier spectra of the time-dependent correlation functions for the dipole-dipole interaction energy. These are
defined as

$$J_{1,2}(t) = \langle Q_{1,2}(0)\,Q_{1,2}(t)\rangle. \tag{192}$$

The random functions Q_1 and Q_2 are given by

$$Q_1(t) = r_{ij}^{-3}\sin\theta_{ij}\cos\theta_{ij}\exp\{-i\varphi_{ij}\}$$
$$Q_2(t) = r_{ij}^{-3}\sin^2\theta_{ij}\exp\{-2i\varphi_{ij}\} \tag{193}$$

where $r_{ij}, \theta_{ij}, \varphi_{ij}$ are the time-dependent polar coordinates defining
$r_{ij}(t)$ with respect to the direction of the vector H_0. An obvious generalization of Eq. (192) is possible for several pairs of nuclei.

In the following, we discuss the meaning of Eq. (191) for a simple
case: We consider a proton as a reference nucleus bound to a rotating
molecule in a solid; we compare this with the corresponding results for
incoherent neutron scattering on such a proton. Let us recall Section 7;
we have considered scattering on an individual proton bound to a
molecule which performs orientational jumps between discrete quasi-
equilibrium *positions:* The probability for an occupancy of a site v at
time t has been called $p_v^{(\mu)}(t)$ assuming that, at $t = 0$, a certain site $r_\mu = 0$
was occupied. From these functions, the self-correlation function G_s

and $S_{inc}(\kappa, \omega)$ have been calculated (Eqs. (144 ff)). As a result, the latter has been obtained as a superposition of Lorentzians.

Here we are concerned with a jump process which occurs between certain discrete *orientations of the vectors* r_{ij} connecting the reference proton with all other protons within the same molecule (intermolecular interactions will be neglected at this point). This leads to a set of discrete values of the quantities Q_1, Q_2 where each orientation has a time-dependent probability of occupancy. From these probabilities one easily calculates the functions $\tilde{J}(t)$ being, in general, exponentials of the form $\langle Q^2(0) \rangle \exp\{-t/\tau\}$ (see for instance [121]); the factors $\langle Q^2(0) \rangle$ depend on the jump geometry. Therefore, $1/T_1$ from (191) is given by a superposition of Lorentzians of the form

$$\tilde{J}(\omega) \sim \frac{\tau}{1 + \omega^2 \tau^2} . \tag{194}$$

This function has a maximum for a value of $\tau = 1/\omega$.

For fast rotational processes ($\omega\tau \ll 1$), the Fourier spectrum is broad and one practically measures the $\omega = 0$-component of the Fourier spectrum, namely $\tilde{J}(0) = 2 \int_0^\infty J(t)\, dt$. In particular, it can be shown that for simple geometries, e.g. for protons in a tetrahedral spherical top molecule, $1/T_1$ can directly be related ot the Raman correlation function as introduced before [191], namely

$$1/T_1 \sim \int_0^\infty \langle P_2[\boldsymbol{u}(0)\, \boldsymbol{u}(t)] \rangle \, dt . \tag{195}$$

The unit vector $\boldsymbol{u}(t)$ connects a *pair* (or any pair) of protons in the molecule (see Table 4). A similar formalism holds for electric nuclear quadrupole interactions [189] which we do not treat here.

We may conclude that the correlation functions $J(t)$ (determining $1/T_1$) and the functions $p_\nu^{(\mu)}(t)$ (entering the incoherent scattering law) are determined by the same rotational jump rate. Therefore, the jump rate obtained from both methods should essentially agree, provided that the correct geometrical model for the rotation has been applied. Also the functions $\langle P_2[\boldsymbol{u}(0)\, \boldsymbol{u}(t)] \rangle$ as found according to (195), and from incoherent neutron scattering should agree in principle.

Such a statement fails if extramolecular interactions (between protons in *adjacent molecules*) are not negligible. This introduces additional time fluctuations in $H(t)$ which are connected with the relative motion of the nuclei in neighbouring molecules. This effect has been discussed for instance in the case of plastic crystals [134]; it gives rise to extramolecular terms in $1/T_1$; they are characterized by a smaller correlation time than that appearing in the intramolecular terms. These extramolecular motional effects tend to complicate the interpretation of nuclear magnetic resonance data, in contrast to incoherent neutron scattering where only the motion of a single proton is observed.

The essential merit of neutrons as compared to nuclear magnetic resonance lies in the fact that very short correlation times τ can be investigated (10^{-9} sec to 10^{-12} sec). For *nmr*, this time region is beyond the minimum of $1/T_1$; consequently the relaxation rate $1/T_1$ is very small ($\omega\tau \ll 1$, see Eq. (194)) and the *nmr* results might be severely affected by spurious interactions, for instance due to paramagnetic impurities. In the case of a distribution of relaxation times, a comparison between nuclear magnetic resonance and neutron scattering would allow to reveal the high frequency components of the distribution.

From *ultrasonic experiments* one obtains the *complex* bulk modulus as a function of sound frequency $\omega/2\pi$ [192, 126] which can be written as

$$K_c(\omega) = K(\omega) + i\omega\eta_v(\omega)$$

$$= K_0 + (i\omega/Vk_BT) \int_0^\infty \exp\{-i\omega t\} \langle\sigma(0)\,\sigma(t)\rangle \, dt. \tag{196}$$

The quantity $K(\omega)$ is the bulk modulus; $K(0)$ is its static value; $\eta_v(\omega)$ is the bulk viscosity which is responsible for the dissipation of energy during sound propagation. $\sigma(t)$ is the time-dependent stress. Assuming that the stress correlation function is

$$\langle\sigma(0)\,\sigma(t)\rangle = \langle\sigma^2\rangle \exp\{-t/\tau_K\} \tag{197}$$

one obtains

$$K_c(\omega) = K_0 + K_r \frac{\omega^2\tau_K^2}{1+\omega^2\tau_K^2} + iK_r \frac{\omega\tau_K}{1+\omega^2\tau_K^2}. \tag{198}$$

Here K_r is the relaxational part of K. A similar relation holds for the complex *dielectric constant;* it is determined by the dipole correlation function $\langle p(0)\,p(t)\rangle = \langle p^2\rangle \exp\{-t/\tau_\varepsilon\}$ which is characterized by a Debye relaxation time τ_ε. In both cases, the mechanical or electric loss has a maximum for $\omega = 1/\tau_K$ or $1/\tau_\varepsilon$.

As has been pointed out before, the rate $1/\tau$, as determined from quasielastic scattering, and from nuclear magnetic resonance should essentially agree, if the proper geometrical model has been applied. The characteristic time of the individual motion, τ, agrees not necessarily with the macroscopic relaxations times τ_K and τ_ε because of the mechanical or electric interaction of the molecules among each other, respectively. In terms of a simple internal field model [193–195] one finds that the relaxation time τ_ε for the collective behaviour of a dipole system is related to τ, the "atomistic relaxation time", by the relation

$$\tau_\varepsilon/\tau = [\varepsilon(0) + 2]/[\varepsilon(\infty) + 2] \tag{199}$$

This leads to a slowing down of the relaxation time if the static dielectric constant $\varepsilon(0)$ increases. A similar factor holds for the ratio τ_K/τ. Therefore, the macroscopic relaxation times are, in general, larger than those determined by quasielastic scattering and by nuclear magnetic resonance.

There is a further complication. So far we have tacitly assumed that the motion of the nucleus responsible for the characteristic time in nuclear magnetic resonance or in neutron scattering is identical with that motion which controls dielectric or acoustic absorption. This is not necessarily the case: A molecule is able to perform different kinds of motions, for instance, if there are different molecular groups rotating independently. This also leads to discrepancies between the "atomistic" time τ (from *nmr* and quasielastic scattering), and τ_ε or τ_K. A detailed discussion of these comparisons yields important informations on the different kinds of molecular motions [194, 195, 126, 139].

11.3 Mössbauer Effect

The cross section for resonance absorption of gamma rays is determined by the self-correlation function $G_s(r, t)$ describing the individual motion of the absorbing nucleus [196]. One obtains, in analogy to Eq. (22b)

$$\sigma(\kappa_0, \omega) = (\sigma_0 \Gamma_\gamma/4\hbar) \int G_s(r, t) \exp\{i(\kappa_0 r - \omega t) - \Gamma_\gamma|t|/2\hbar\} \, dr \, dt \tag{200}$$

Here, σ_0 is the cross section if the energy of the incident radiation $\hbar\omega'$ coincides with the resonance energy of the absorber at $\hbar\omega_0$. Furthermore, Γ_γ is the *natural width* of the excited state of the nucleus, and $\hbar\omega = \hbar(\omega_0 - \omega')$ the energy shift. The momentum of the interacting quanta is called $\hbar\kappa_0$.

For an absorbing atom with a finite mobility, the Mössbauer absorption peak described by $\sigma(\kappa_0, \omega)$ is broadened by the diffusive motion as well as by the nuclear life time. Obviously, such a diffusion broadening can only be observed if it is of the same order of magnitude as Γ_γ, or, with other words, if the natural nuclear life-time $\hbar/\Gamma_\gamma$ is comparable with the characteristic time of the motion. Under certain conditions not to be discussed here, the resulting width of the Mössbauer line is then a sum of the natural width Γ_γ and the width due to the diffusive motion. Because of the extreme smallness of Γ_γ, quasielastic Mössbauer absorption has to be applied to the investigation of very slow diffusive motions which are normally not accessible to neutron spectroscopy. This happens, for instance, for viscous media [148, 197] or with problems concerning the atomic diffusion mechanism in solids [198, 199].

As can be seen from Eq. (200), Mössbauer experiments should give, in principle, the same physical information as quasielastic neutron scattering. Besides the possibility of studying very slow motions, an important difference lies in the fact that the momentum transfer $\hbar\kappa_0$ is quite *large* $(\kappa^2 D\tau_r \gg 1)$. This needs modifications of the theory (as discussed in Section 4) with regard to the proper description of the short-time behaviour of $G_s(r, t)$ [197].

Acknowledgement. The author is very indebted to Professor H. Maier-Leibnitz and Dr. B. Jacrot for the kind hospitality in the Institute von-Laue-Langevin where he was able to write this article, and for many stimulating discussions on this subject. The author further acknowledges comments, discussions and helpful criticism from many colleagues, Professor K.-E. Larsson in particular, and from Dr. H. Stiller, Dr. K. Michel, Dr. F. Hossfeld, Dr. R. R. Lechner and Dr. H. Lütgemeier.

Note Added in Proof. In the Fifth International IAEA Symposium on Inelastic Neutron Scattering (Grenoble, March 1972) a considerable amount of work has been presented concerning applications of quasielastic scattering (proceedings in print by the International Atomic Energy Agency, Vienna). A number of experimental papers was dealing with molecular rotations in solids and in liquids. Practically, they all follow the concepts of theoretical interpretation as outlined in Sections 7 and 8. Only three papers were treating polymeric motions, in particular a theoretical one giving a generalization of the theories mentioned in 9.1 (IV). In general it turned out that there is a strong tendency to work on chemical applications of the method. Careful experiments on S_{coh} were dealing with simple liquids in the nonhydrodynamic region (see Section 10.2). Several interesting papers on quasielastic scattering for the study of hydrogen diffusion in metals will be published in the proceedings of the International Meeting on Hydrogen in Metals (Jülich, March 1972, see JÜL-Conf-6, Vol. 1, p. 259 ff.).

References

1. Turchin, V. F.: Slow neutrons; Jerusalem: Israel Program for Sci. Transl. 1965.
2. Gurevich, I. I., Tarasov, L. V.: Low energy neutron physics. Amsterdam: North-Holland Publ. Comp. 1968.
3. Egelstaff, P. A. (Ed.): Thermal neutron scattering. London: Academic Press 1965.
4. Neutron inelastic scattering; International Atomic Energy Agency, Conference proceedings: Vienna 1961 (Vienna Symposium); Vienna 1963 (Chalk River Symposium); Vienna 1965 (Bombay Symposium); Vienna 1968 (Copenhagen Symposium), see also: Panel Conference on Instrumentation for Neutron Inelastic Scattering, International Atomic Energy Agency, Vienna 1969.
5. van Hove, L.: Phys. Rev. **95**, 249 (1954).
6. Messiah, A.: Quantum mechanics. Amsterdam: North-Holland Publ. Comp. 1961 and 1962.
7. Fermi, E.: Ric. Sci. Vicotruiz. **7**, part 2, 13 (1936).
8. Sarma, G.: Neutron inelastic scattering. Vienna, Inernational Atomic Energy Agency 1961, p. 397 and Ref. [3], p. 437.
9. Herzberg, G.: Molecular spectra. New York: 1945; see also Stiller, H., Hautecler, S.: Z. Physik **166**, 393 (1962).
10. Lippmann, B. A.: Phys. Rev. **79**, 481 (1950).

11. Summerfield, G. C.: Ann. Phys. **26**, 72 (1964), and Stassis, C.: Phys. Rev. Letters **24**, 1415 (1970).
12. Grimm, H., Stiller, H., Plesser, Th.: Phys. Stat. Sol. **42**, 207 (1970).
13. van Hove, L.: Physica **24**, 404 (1958).
14. Rahman, A.: Phys. Rev. **136**, A 405 (1964).
15. — Singwi, K. S., Sjölander, A.: Phys. Rev. **126**, 986 (1962).
16. Vineyard, G. H.: Phys. Rev. **110**, 999 (1958).
17. Nijboer, B. R. A., Rahman, A.: Physica **32**, 415 (1966).
18. Schofield, P.: Neutron inelastic scattering. Vienna: International Atomic Energy Agency 1961, p. 39, and Desai, R. C., Nelkin, M. S.: Nucl. Sci. Eng. **24**, 142 (1966).
19. Harp, G. D., Berne, B. J.: Phys. Rev. A **2**, 975 (1970).
20. Nelkin, M.: Neutron inelastic scattering, p. 3. Vienna: International Atomic Energy Agency 1961.
21. Becker, R.: Theorie der Wärme. Berlin-Heidelberg-New York: Springer 1966
22. Dicke, R. H.: Phys. Rev. **89**, 472 (1953).
23. de Gennes, P. G.: Physica **25**, 825 (1959).
24. Placzek, G.: Phys. Rev. **86**, 377 (1952).
25. Randolph, P. D.: Phys. Rev. **134** A, 1238 (1964).
26. Maier-Leibnitz, H.: Nukleonik **8**, 61 (1966).
27. Kalus, J.: Unpublished.
28. Alefeld, B., Birr, M., Heidemann, A.: Naturwissenschaften **56**, 410 (1969).
29. — To be published (1971).
30. Gompf, F., Reichardt, W., Gläser, W., Beckurts, K. H.: Neutron inelastic scattering, Vol. 2, p. 417. Vienna: International Atomic Energy Agency 1968.
31. Pál, L., Kroó, N., Pellionisz, P., Szlávik, F., Vizi, I.: Neutron inelastic scattering, Vol. 2, p. 407. Vienna: International Atomic Energy Agency 1968.
32. Sköld, K.: Nucl. Instr. Meth. **63**, 114 and 347 (1968).
33. Hossfeld, F., Amadori, R., Scherm, R.: Proc. of the Panel Conference on Instrumentation for Neutron Inelastic Scattering, p. 117. Vienna: International Atomic Energy Agency 1970.
34. Cooper, M. J., Nathans, R.: Acta Cryst. **23**, 357 (1967).
35. Melkonian, E.: Proc. Int. Conference on Peaceful Uses of Atomic Energy, Vol. 4, p. 340. New York: United Nations 1956.
36. Schelten, J., Hossfeld, F.: J. Appl. Cryst. **4**, 210 (1971).
37. Cocking, S. J.: J. Phys. C., (Sol. Stat. Phys.) **2**, 2047 (1969).
38. Vineyard, G. H.: Phys. Rev. **96**, 93 (1954).
39. White, J. W.: J. Macromolec. Sci. Chem. A **4**, 1275 (1970).
40. Aldred, B. K., Eden, R. C., White, J. W.: Disc. Faraday Soc. **43**, 169 (1967).
41. Hughes, D. J., Schwartz, R. B.: Neutron cross sections, BNL – 325, United States Atomic Energy Comm. 1955, and Data Compilation of C.I.N.D.A.
42. Andriesse, C. D., Compagner, A., Hasman, A., van Loef, J., van Zevenbergen, F.: Phys. Letters **28** A, 642 (1969).
43. Sköld, K.: To be published (1971), see [178].
44. Egelstaff, P. A.: An introduction to the liquid state. London: Academic Press 1967.
45. Larsson, K. E.: Neutron inelastic scattering, Vol. 1, p. 397. Vienna: International Atomic Energy Agency 1968.
46. Chandrasekhar, E.: Rev. Mod. Phys. **15**, 1 and 8 (1943).
47. Glass, L., Rice, S. A.: Phys. Rev. **176**, 239 (1968).
48. Brockhouse, B. N.: Nuovo Cimento **9**, (Suppl. 1), 45 (1958).
49. Cribier, D., Jacrot, B.: J. Phys. **21**, 69 (1960).
50. Hughes, D. J., Palewsky, H., Kley, W., Tunkelo, E.: Phys. Rev. **119**, 872 (1960).

51. Larsson, K. E., Dahlborg, U., Holmryd, S.: Ark. Fys. **17**, 369 (1960), and Larsson, K. E., Dahlborg, U., Reactor. Sci. Techn. (J. Nucl. Eng. A/B) **16**, 81 (1962).
52. Egelstaff, P. A., Schofield, P.: Nucl. Sci. Eng. **12**, 260 (1962).
53. — Advan. Phys. **11**, 203 (1962).
54. Rahman, A., Singwi, K. S., Sjölander, A.: Phys. Rev. **126**, 997 (1962).
55. Sears, V. F.: Proc. Phys. Soc. **86**, 953 (1965).
56. Damle, P. S., Sjölander, A., Singwi, K. S.: Phys. Rev. **165**, 277 (1968), see also: Nakahara, Y., Takahashi, H.: Proc. Phys. Soc. **88**, 747 (1966).
57. Caroli, B., Jannink, G., Saint-James, D.: J. Phys. C. (Sol. Stat. Phys.) **4**, 545 (1971).
58. Sjölander, A.: Arkiv Fysik **14**, 315 (1958).
59. Zwanzig, R.: Lectures in theoretical physics. In: Britlin, W. B., Downs, W. B., Downs, J., Kubo, R. (Eds.): Tokyo: Syokabo and New York: Benjamin 1966.
60. Mori, H.: Progr. Theor. Phys. (Kyoto) **33**, 423 (1965) and **34**, 399 (1965).
61. Berne, B. J., Boon, J. P., Rice, S.: J. Phys. Chem. **45**, 1086 (1966).
62. Singwi, K. S., Tosi, M. P.: Phys. Rev. **157**, 153 (1967).
63. Desai, R. C.: Phys. Rev. A **3**, 320 (1971).
64. Akcasu, A. Z., Daniels, E.: Phys. Rev. A **2**, 962 (1970).
65. Jain, S. C., Bhandari, R. C.: Physica **52**, 393 (1971).
66. Singwi, K. S., Sjölander, A.: Phys. Rev. **167**, 152 (1968).
67. Bonamy, L., Galatry, L.: Physica **46**, 133 (1970).
68. Rahman, A.: Riv. Nuovo Cimento **1** (numero speciale) 315 (1969).
69. Verlet, L.: Phys. Rev. **159**, 98 (1967) and **165**, 205 (1968).
70. Levesque, D., Verlet, L.: Phys. Rev. A **2**, 2514 (1970).
71. Schiff, D.: Phys. Rev. **186**, 151 (1969).
72. See Nolting, W.: Z. Physik **242**, 199 (1971).
73. Harp, G. D., Berne, B. J.: J. Phys. Chem. **49**, 1249 (1968).
74. Alder, B. J., Wainwright, T. E.: Phys. Rev. A **1**, 18 (1970). Wainwright, T. E., Alder, B. J., Gass, D. M.: Phys. Rev. A **4**, 233 (1971).
75. Zandveld, P., Andriesse, C. D., Bregman, J. D., Hasman, A., van Loef, J. J.: Physica **50**, 511 (1970).
76. Sköld, K., Larsson, K. E.: Phys. Rev. **161**, 102 (1967), see also Chen, S. H., Eder, O. J., Egelstaff, P. A., Haywood, B. C. G., Webb, F. J.: Phys. Letters **19**, 269 (1965).
77. Dasannacharya, B. A., Rao, K. R.: Phys. Rev. **137**, A 417 (1965) and Brockhouse, B. N., Bergsma, J., Dassannacharya, B. A., Pope, N. K.: Neutron inelastic scattering, Vol. 1, p. 189. Vienna: International Atomic Energy Agency 1963.
78. Zwanzig, R., Bixon, M.: Phys. Rev. A **2**, 2005 (1970).
79. Andriesse, C. D.: Phys. Letters **33** A, 419 (1970).
80. Cabane, B., Friedel, J.: J. Phys. **32**, 73 (1971).
81. Frenkel, J.: Kinetic theory of liquids. London: Oxford Univ. Press 1946.
82. Singwi, K. S., Sjölander, A.: Phys. Rev. **119**, 863 (1960).
83. Oskotskii, V. S.: Soviet Phys.-Solid State **5**, 789 (1963).
84. Stiller, H., Danner, H. R.: Neutron inelastic scattering, p. 363. Vienna: International Atomic Energy Agency 1961.
85. Sakamoto, M., Brockhouse. B. N., Johnson, R. G., Pope, N. K.: J. Phys. Soc. Japan **17**, Suppl. B-II, 370 (1962), see also [52].
86. Brugger, R. M.: Nucl. Sci. Eng. **33**, 187 (1968).
87. Safford, G. J., Schaffer, P. C., Leung, P. S., Doebbler, G. F., Brady, G. W., Lynden, E. F. X.: J. Chem. Phys. **50**, 2140 (1969).
88. Franks, F., Ravenhill, J., Egelstaff, P. A., Page, D. I.: Proc. Roy. Soc. (London) A **319**, 189 (1970).
89. Harling, O. K.: J. Chem. Phys. **50**, 5279 (1969), see also Nelkin, M.: Phys. Rev. **119**, 741 (1960).

90. Eucken, A.: Nachr. Akad. Wiss. Göttingen, Math. Phys. Kl. **2**, 38 (1946).
91. White, J. W.: Proc. Dtsch. Bunsenges. (Herrenalb Meeting 1971), in print.
92. Hertz, H. G., Zeidler, M. D.: Ber. Bunsenges. Phys. Chem. **67**, 774 (1963).
93. Bloembergen, N., Purcell, E. M., Pound, R. V.: Phys. Rev. **73**, 679 (1948).
94. Springer, T.: Nukleonik **3**, 110 (1961) (review).
95. Harling, O. K.: Nucl. Sci. Eng. **33**, 41 (1968).
96. Whittemore, W. L.: Nucl. Sci. Eng. **33**, 195 (1968).
97. Alefeld, G.: Phys. stat. sol. **32**, 67 (1969), for a general review see Gibb, T. R. P.: Progr. Inorgan. Chem. (F. A. Cotton, Ed.), Vol. 3, p. 315. New York: Interscience Publ. 1962.
98. Chudley, C. T., Elliott, R. J.: Proc. Phys. Soc. **77**, 353 (1961).
99. Blaesser, G., Perretti, J.: Proc. Int. Conference on Vacancies and Interstitials in Metals, KFA Jülich, Jül-Conf-2, Vol. **2**, p. 886 (1968).
100. Rowe, J. M., Sköld, K., Flotow, H. E., Rush, J. J.: J. Phys. Chem. Solids **32**, 41 (1971), see also Rush, J. J., Flotow, H. E.: J. Chem. Phys. **48**, 3795 (1968).
101. Gissler, W., Rother, H.: Physica **50**, 380 (1970).
102. Stiller, H., Springer, T.: Z. Naturforsch. **26 a**, 575 (1971).
103. Kaufmann, B., Lipkin, H. J.: Ann. Phys. **18**, 294 (1962).
104. Kley, W.: Z. Naturforsch. **21 a**, 1770 (1966). (Supplement).
105. Krivoglaz, M. A.: Soviet Phys. JETP **19**, 432 (1964).
106. Dawber, P. G., Elliott, R. J.: Proc. Roy. Soc. London A **273**, 222 (1963), see also: Kagan, Y., Iosilevskii, Y., Soviet. Phys. JETP **17**, 925 (1963).
107. Blaesser, G., Perretti, J., Toth, G.: Phys. Rev. **171**, 665 (1968).
108. Sköld, K., Nelin, G.: J. Phys. Chem. Solids **28**, 2369 (1967); see also Nelin, G.: Phys. Stat. Sol. (b) **45**, 527 (1971).
109. Beg, M. M., Ross, D. K.: J. Phys. C.: Solid Stat. Phys. **3**, 2487 (1970).
110. Verdan, G., Rubin, R., Kley, W.: Neutron inelastic scattering, Vol. 1, p. 223. Vienna: International Atomic Energy Agency 1968.
111. Gissler, W., Alefeld, G., Springer, T.: J. Phys. Chem. **31**, 2361 (1970) and Rubin, R., Claessen, Y.: Solid State Commun. **8**, 1321 (1970).
112. Schaumann, G., Völkl, J., Alefeld, G.: J. Phys. Chem. Solids **31**, 1805 (1970).
113. Zamir, D., Cotz, R. M.: Phys. Rev. **134** A, 666 (1964).
114. Arons, R. R., Tamminga, Y., de Vries, G.: Phys. Stat. Sol. **40**, 107 (1970).
115. Flynn, C. P., Stoneham, A. M.: Phys. Rev. B **1**, 3966 (1970) and Lepski, D.: Phys. Stat. Sol. **35**, 697 (1969).
116. Andrew, A. F., Lifshitz, I. M.: Soviet Phys. JETP **29**, 1107 (1969).
117. Willis, B. T. M., Pawley, G. S.: Acta Cryst. A **26**, 254 (1970) and Pawley, G. S.: Phys. Stat. Sol. **20**, 347 (1967).
118. Brot, C., Renaud, M.: Bull. Soc. Cryst. France, (1971) to be published. — Darmon, I., Brot, C.: Molecular Crystals **2**, 301 (1967).
119. J. Phys. Chem. Solids **18**, 1 ff. (1961); J. Chim. Phys. **63**, 1 ff.
120. Mertens, F., Biem, W., Hahn, H.: Z. Physik **220**, 1 (1969) and Hardy, W., Silvera, J., McTague, J.: Phys. Rev. Letters **22**, 297 (1969).
121. Itoh, J., Kasaka, R., Saito, Y.: J. Phys. Soc. Japan **17**, 463 (1962), and Itoh, J., Yamagata, Y.: J. Phys. Soc. Japan **17**, 481 (1962).
122. Stockmeyer, R., Stiller, H.: Phys. Stat. Sol. **27**, 269 (1968).
123. Sköld, K.: J. Chem. Phys. **49**, 2443 (1968).
124. Sears, V. F.: Can. J. Phys. **45**, 237 (1967), Sears, V. F.: Can. J. Phys. **44**, 1279 (1966).
125. Steele, W. A., Pecora, R.: J. Chem. Phys. **42**, 1863 (1965).
126. Montrose, C. J., Litovitz, T. A.: Neutron inelastic scattering, Vol. 1, p. 623. Vienna: International Atomic Energy Agency 1968.

127. Egelstaff,P.A.: J. Chem. Phys. **53**, 2590 (1970).
128. Agraval,A.K., Yip,S.: Nucl. Sci. Eng. **37**, 368 (1969), see also Phys. Rev. **171**, 263 (1968).
129. Steele,W.A.: J. Chem. Phys. **38**, 2404 and 2411 (1963).
130. Ivanov,E.N.: Soviet Phys. JETP **18**, 1041 (1964).
131. Dahlborg,U., Larsson,K.E., Pirkmajer,E.: Physica **49**, 1 (1970), see [131a]
131a. Larsson,K.E.: Phys. Rev. A **2**, 810 (1971) (erratum).
132. Kapulla,H.: (1971) to be published.
133. Press,W.: J. Phys. Chem. 1972 (in print).
134. Resing,H.A.: Molec. Liquid Cryst. **9**, 101 (1969).
135. Lechner,R.E., Rowe,J.M., Sköld, K., Rush,J.: J. Chem. Phys. Letters **4**, 444 (1969).
136. de Graaf,L.A., Sciensinski,J.: Physica **48**, 79 (1970).
137. Kim,H.J., Goyal,P.S., Venkataraman,G., Dasannacharya,B.A., Thaper,C.L.: Solid State Commun. **8**, 889 (1970).
138. Larsson,K.E.: Phys. Rev. A **3**, 1006 (1971); Larsson,K.E., Bergstedt, L.: Phys. Rev. **151**, 117 (1966), and Larsson,K.E., Amaral,L., Ivanchev,N., Ripeanu,S., Bergstedt,L., Dahlborg,U.: Phys. Rev. **151**, 126 (1966), see [131a].
139. Larsson,K.E.: Phys. Rev. **167**, 171 (1968).
140. Dasannacharya,B.A., Venkataraman,G.: Phys. Rev. **156**, 196 (1967).
141. Harker,Y.D., Brugger,R.M.: J. Chem. Phys. **42**, 275 (1965).
142. Egelstaff,P.A., Haywood,B.C., Webb,F.J.: Proc. Phys. Soc. (London) **90**, 681 (1967); Talhank: J. Chem. Phys. **48**, 1273 (1968).
143. Eder,O.J., Chen,S.H., Egelstaff,P.A.: Proc. Phys. Soc. (London) **89**, 833 (1966), and Eder,O.J., Egelstaff,P.A.: Neutron inelastic scattering, Vol. 2, p. 223. Vienna: International Atomic Energy Agency 1968, and Eder,O.J.: Acta Physica **33**, 181 (1971).
144. Webb,F.J.: Proc. Phys. Soc. (London) **92**, 912 (1967); see also Lurie, N.A., Summerfield,G.C., J. Chem. Phys. **49**, 490 (1968) (oxygen).
145. Agraval,A.K., Yip,S.: Phys. Rev. A **1**, 970 (1970).
146. Larsson, K.E., Dahlborg, U.: Physica **30**, 1561 (1964).
147. Birr,M.: Z. Physik **238**, 221 (1970).
148. Craig,P.P., Sutin,N.: Phys. Rev. Letters **11**, 460 (1963).
149. de Graaf,L.A.: Physica **40**, 497 (1969).
150. Longster,G.F.: White,J.W.: Molec. Phys. **17**, 1 (1969).
151. Sampson,T.E., Carpenter,J.M.: J. Chem. Phys. **51**, 5543 (1969).
152. Neutron inelastic scattering, Vol. 1, p. 475ff., and Vol. 2, p. 143ff. Vienna: International Atomic Energy Agency 1968.
153. Safford,G.J., Naumann,A.W.: Advan. Polymer. Sci. **5**, 1 (1967) and Trevino,S., Boutin,H.: J. Macromol. Sci. (Chem.) A **1**, 723.
154. Pechold,W.: Kolloid Z. und Z. Polymere **228**, 1 (1968).
155. Rouse,P.E.: J. Chem. Phys. **21**, 1272 (1953).
156. de Gennes,P.G.: Physics **3**, 37 (1967).
157. Zimm,B.H.: J. Chem. Phys. **24**, 269 (1956).
158. Jannink,G., de Gennes,P.G.: J. Chem. Phys. **48**, 2260 (1968).
159. Dubois-Violette,E., de Gennes,P.G.: Physics **3**, 181 (1967); Bidaux,R., Cotton,J.P., Farnoux,B., Jannink,G.: J. Phys. Chem. **54**, 3717 (1971).
160. Axmann,A., Gissler,W., Kollmar,A., Springer,T.: Disc. Faraday Soc. **50**, 74 (1970); Kollmar,A., (Thesis Aachen) 1971.
161. Ring,J.W., Egelstaff,P.A.: J. Chem. Phys. **51**, 762 (1969).
162. Janik,J.A., Janik,J.M., Otnes,K., Riste,T.: (1971) to be published.
163. Blinc,R., Dimic, V., Pirs,J., Vilfan,M., Zupancic,I.: Molec. Cryst. and Liquid Cryst. **14**, 97 (1971).

164. Leung,P.S., Safford,G.J.: J. Phys. Chem. **74**, 3696 (1970).
165. Safford,G.J., Leung,P.S., Naumann,A.W., Schaffer,P.C.: J. Chem. Phys. **50**, 4444 (1969).
166. Egelstaff,P.A., Stretton Downes,J., White,J.W.: In: Molecular sieves. The Society for Chemistry and Industry (publisher), 1966, p. 306.
167. Leung,P.S., Sanborn,S.M., Safford,G.J.: J. Phys. Chem. **74**, 3710 (1970).
168. Anderson,R.G.W., White,J.W.: Disc. of the Faraday Society on Thin Liquid Films and Boundary Layers, 1971 (in print) and Olejnik,S., Stirling,G.C., White, J.W., 1971 (in print).
169. Verdan,G.: Phys. Letters **25**A, 436 (1967).
170. Tait,M.J., Franks,F.: Nature **230**, 91 (1971).
171. Landau,L.D., Lifshitz,E.M.: Electrodynamics of continuous media, Reading (Mass.) Addison-Wesley Comp. 1960, and Fluid Mechanics, London, Pergamon Press 1959.
172. Pecora,R.: J. Chem. Phys. **40**, 1604 (1964).
173. Mountain,R.D.: Rev. Mod. Phys. **38**, 205 (1966).
174. Kadanoff,L.P., Martin,P.C.: Ann. Phys. **24**, 419 (1963).
175. Singwi,K.S.: Physica **31**, 1257 (1965).
176. — Phys. Rev. **136**A, 969 (1964).
177. Glass,L., Rice,S.A.: Phys. Rev. **165**, 186 (1968).
178. Singwi,K.S., Sköld,K., Tosi,M.P.: Phys. Rev. A **1**, 454 (1970) and Pathak,K.N., Singwi,K.S.: Phys. Rev. A **2**, 2427 (1970).
179. Jain,S.C., Saxena,N.S., Bhandari,R.C.: J. Chem. Phys. **52**, 4629 (1970).
180. Kerr,W.C.: Phys. Rev. **174**, 316 (1968) and Sears,V.F.: Can. J. Phys. **47**, 199 (1969); **48**, 616 (1970); Phys. Rev. **185**, 200 (1969).
181. Chung,Ch., Yip,S.: Phys. Rev. **182**, 323 (1969).
182. Desai,R.C., Yip,S.: Phys. Rev. **180**, 299 (1969).
183. Andriesse,C.D.: Physica **48**, 61 (1970) and **49**, 502 (1970).
184. Venkataraman,G., Sahni,V.C.: Rev. Mod. Phys. **42**, 409 (1970).
185. Egelstaff,P.A., Page,D.I., Powles,J.G.: Molec. Phys. **20**, 881 (1971).
186. Schofield,P.: Proc. Phys. Soc. (London) **88**, 149 (1966).
187. Gordon,R.G.: J. Chem. Phys. **43**, 1307 (1965).
188. — J. Chem. Phys. **44**, 1830 (1966).
189. Abragam,A.: The principles of nuclear magnetism. Oxford: Oxford University Press 1961.
190. Powles,J.P.: Neutron inelastic scattering, Vol. 1, p. 379. Vienna: International Atomic Energy Agency 1968.
191. Gordon,R.G.: J. Chem. Phys. **42**, 3658 (1965), Gordon,R.G.: In: Waugh,J.S. (Ed.): Advances in magnetic resonance, Vol. 3, p. 1. New York: Academic Press 1968.
192. Montrose,C.J., Solovejev,V.A., Litovitz,T.A.: J. Acoust. Soc. Am. **43**, 117 (1968), and Pinnow,D.A., Candau,S.J., LaMacchia,J.T., Litovitz,T.A.: J. Acoust. Soc. Am. **43**, 131 (1968).
193. Debye,P.: Polar molecules. New York: Dover Publ. Inc. 1945. and Powles,J.G.: J. Chem. Phys. **21**, 633 (1953).
194. Litovitz,T.A., Sette,D.: J. Chem. Phys. **21**, 17 (1953).
195. Luszczynski,K., Kail,J.A.E., Powles,J.G.: Proc. Phys. Soc. (London) **75**, 243 (1960).
196. Singwi,K.S., Sjölander,A.: Phys. Rev. **120**, 1093 (1960), and Singwi,K.S.: Neutron inelastic scattering, Vol. 1, p. 3. Vienna: International Atomic Energy Agency 1963.
197. Bhide,V.G., Sundaram,R., Bhasin,H.C., Bonchev,T.: Phys. Rev. B **3**, 673 (1971).
198. Boyle,A.J.F., Bunbury,D., Edwards,C., Hall,H.E.: Proc. Phys. Soc. **77**, 129 (1961a).
199. Knauer,R.C.: Phys. Rev. B **3**, 567 (1971).

200. Schneider, T.: Phys. Rev. A 3, 2145 (1971), and Schneider, T., Brout, R., Thomas, H., Feder, J.: Phys. Rev. Letters **25**, 1423 (1970).
201. Larsson, K. E.: Personal comm.

Professor Dr. Tasso Springer
Institut für Festkörperforschung
der Kernforschungsanlage Jülich GmbH
D-5170 Jülich, Germany

Springer Tracts in Modern Physics

To appear in forthcoming volumes:

H. Arenhövel and H. J. Weber: Nuclear Isobar Configurations

J. Brandmüller and R. Claus: Light Scattering on Optical Phonons and Polaritons

R. Graham: Statistical Theory of Instabilities in Stationary Non-Equilibrium Systems with Applications to Lasers and Non-Linear Optics

K. Heinloth: Experiments on Electroproduction in High Energy Physics

D. Schmid: Nuclear Magnetic Double Resonance – Principles and Applications in Solid State Physics

H. Theissen: Spectroscopy of Light Nuclei by Low Energy (70 MeV) Inelastic Electron Scattering

Volume 63

Photon-Hadron Interactions II. Lectures presented at the International Summer Institute in Theoretical Physics. DESY, Hamburg, July 12—24, 1971

A. P. Contogouris: Regge Analysis and Dual Absorptive Model

A. Donnachie: Exotic Electromagnetic Currents

J. Frøyland: High Energy Photoproduction of Pseudoscalar Mesons

F. M. Renard: p-ω Mixing

D. Schildknecht: Vector Meson Dominance, Photo- and Electro-production from Nucleons

K. Schilling: Some Aspects of Vector Meson Photoproduction on Protons

P. D. B. Collins and F. D. Gault: The Eikonal Model for Regge Cuts in Pion-Nucleon Scattering

Volume 62

With 46 figures
V, 147 pages. 1972
Cloth DM 58,—

Photon Hadron Interactions I. International Summer Institute in Theoretical Physics. DESY, July 12—24, 1971

R. Jackiw: Canonical Light-Cone Commutators and Their Applications
H. D. Dahmen: Local Saturation of Commutator Matrix Elements
P. V. Landshoff: Duality in Deep Inelastic Electroproduction
C. H. Llewellyn Smith: Parton Models of Inelastic Lepton Scattering
H. R. Rubinstein: Duality for Real and Virtual Photons
V. Rittenberg: Scaling in Deep Inelastic Scattering with Fixed Final States
K. Huang: Duality and the Pion Electromagnetic Form Factor
K. Huang: Deep Inelastic Hadronic Scattering in Dual-Resonance Model
G. Furlan, N. Paver, and C. Verzegnassi: Low Energy Theorems and Photo- and Electroproduction near Threshold by Current Algebra

Volume 61

With 41 figures
IV, 166 pages. 1972
Cloth DM 68,—

J. L. Basdevant: $\pi\pi$ Theories
A. Donnachie: The Nucleon Resonances
G. Gustafson and J. Hamilton: The Dynamics of Some πN Resonances
B. Schrempp-Otto and F. Schrempp: Are Regge Cuts Still Worthwhile?
R. Oehme: Rising Cross-Sections
B. Renner: On the Problem of the Sigma Terms in Meson-Baryon Scattering
H. Genz: Local Properties of σ-Terms: A Review
P. Brinckmann: Polarization of Recoil Nucleons from Single Pion Photoproduction

Distributor in USA:
Springer-Verlag New York, Inc.
175 Fifth Ave, New York, N. Y. 10010